A Short Guide to Writing about Chemistry

THE SHORT GUIDE SERIES
Under the Editorship of
Sylvan Barnet
Marcia Stubbs

A *Short Guide to Writing about Literature* by Sylvan Barnet

A *Short Guide to Writing about Art* by Sylvan Barnet

A *Short Guide to Writing about Biology* by Jan A. Pechenik

A *Short Guide to Writing about Social Science* by Lee J. Cuba

A *Short Guide to Writing about Film* by Timothy Corrigan

A *Short Guide to Writing about History* by Richard Marius

A *Short Guide to Writing about Science* by David Porush

A *Short Guide to Writing about Chemistry*
by Herbert Beall and John Trimbur

A Short Guide to Writing about Chemistry

HERBERT BEALL
Department of Chemistry
Worcester Polytechnic Institute

JOHN TRIMBUR
Department of Humanities and Arts
Worcester Polytechnic Institute

HarperCollins*CollegePublishers*

Senior Editor: Patricia Rossi
Project Coordination and Text Design: Ruttle, Shaw & Wetherill, Inc.
Cover Designer: Kay Petronio
Cover Photograph: Ted Horowitz/The Stock Market
Electronic Production Manager: Christine Pearson
Manufacturing Manager: Helene G. Landers
Electronic Page Makeup: Ruttle, Shaw & Wetherill, Inc.
Printer and Binder: R. R. Donnelley and Sons Company
Cover Printer: Phoenix Color Corp.

A Short Guide to Writing about Chemistry

Library of Congress Cataloging-in-Publication Data

Beall, Herbert.
A short guide to writing about chemistry / Herbert Beall, John Trimbur.
p. cm. — (The short guide series)
Includes index.
ISBN 0-673-46882-8 (pbk.)
1. Chemistry—Authorship. 2. Communication in chemistry.
I. Trimbur, John. II. Title.
QD9.15.B43 1996
540'.7—dc20 95-46045
CIP

96 97 98 9 8 7 6 5 4 3 2

Contents

PREFACE ix

1—WRITING AND CHEMISTRY 1

What Do Chemists Read and Write about? 2

How Writing about Chemistry Can Help You Become a Better Writer 5

2—WRITING AND SCIENTIFIC RESPONSIBILITY 12

Scientific Honesty 12

Understanding the Competitive World of Science 14

Reading: James D. Watson, from The Double Helix *15*

A Case of Possible Dishonesty 17

Readings: David Baltimore, "Baltimore's Travels" 18

Pamela Zurer, "Scientific Whistleblower Vindicated" 29

David Weaver, et al., "Retraction," Cell *33*

3—READING AND WRITING TO LEARN CHEMISTRY 34

Why Does Chemistry Seem Difficult to So Many Students? 34

Reading Your Textbook 37

Taking and Using Lecture Notes 44

Studying for Quizzes and Examination 46

4—WRITING LABORATORY REPORTS 52

The Laboratory in the Study of Chemistry 52

Reading: Frederick H. Getman, from Life of Ira Remsen *53*

Preparing for the Laboratory 54

The Laboratory Notebook 55

The Laboratory Report 70

5—HOW TO READ A SCIENTIFIC ARTICLE: WRITING SUMMARIES AND CRITIQUES 90

Understanding Scientific Articles 91

Reading Scientific Articles 92

Reading: James D. Watson and Francis H. C. Crick, "A Structure for Deoxyribose Nucleic Acid" 93

Writing Summaries 98

Writing Critiques 99

Reading: Michael E. Potts and Duane R. Churchwell "Removal of Radionuclides in Wastewaters Utilizing Potassium Ferrate(VI)" 100

6—WRITING LITERATURE REVIEWS 109

Interpretation in Chemistry 110

Readings: Thomas Henry Huxley, "Joseph Priestley" 111

W. L. Bragg, "The Structure of Some Crystals" 114

W. C. Bray and G. E. K. Branch, "Valence and Tautomerism" 117

G. N. Lewis, "The Atom and the Molecule" 118

Writing the Literature Review 121

Searching the Literature of Chemistry 134

7—WRITING RESEARCH PROPOSALS 143

An Overview of Proposal Writing 144

Writing a Research Proposal 150

8—CHEMISTRY AND THE PUBLIC: WRITING TO INFORM AND PERSUADE 161

Reporting Science: Writing to Inform 161

Reading: "Chemists Dissect the Colors of Flowers" 162

Chemistry and "Chemicals" 164

Chemistry and Public Opinion: Writing to Persuade 166

Readings: Rachel Carson, from Silent Spring *168*

Robert Ostmann, Jr., from Acid Rain: A Plague Upon the Waters *173*

Michael Gough, from Dioxin, Agent Orange: The Facts *177*

CREDITS 183

INDEX 185

Preface

Chemistry is many things to different people. It is the science that is indispensable to producing the food we eat, the clothes we wear, the cars we drive, and the medications we need. It is a career and a way of life to the chemist in industry or government and to the chemistry teachers and professors who teach at essentially every high school, college, and university in the world. And it is a subject of complexity and difficulty for many high school and college students.

The subject of chemistry is matter—anything in the universe that has mass or weight and takes up space. Chemistry involves understanding what matter consists of, right down to the atoms that compose it, and using this understanding to change matter from one form to another in order to produce the chemical products we use in virtually every aspect of our lives.

But chemistry also involves communication—communication between the teacher and the student, communication among students, and communication among teachers and researchers throughout the world. Communication in chemistry takes many forms, such as conversations, prepared talks or lectures, handwritten laboratory records, casual notes, and formal papers and books that are edited and checked many times before they are printed.

Communication is a key to understanding the subject of chemistry. Nobody, either student or experienced researcher, is going to advance or learn about chemistry unless the channels of communication are open and he or she knows how to use them.

THE PURPOSE OF THIS BOOK

This book is about communication in chemistry, particularly reading and writing. Its purpose is to introduce chemistry students to the ways chemists communicate with each other, with students, and with the public. The information in this book will be useful to people at many levels who are studying or practicing chemistry. The book covers the most important forms of communication in chemistry—textbooks, lectures, laboratory reports, articles on research results, literature reviews, research proposals, and science writing for the public. It offers both beginning and more advanced chemistry students practical advice on how to read and write about chemistry.

THE DESIGN OF THIS BOOK

The design of this book follows the progress of a chemistry student, from introductory level coursework in general chemistry to advanced work in more specialized aspects of chemistry.

The first two chapters cover issues that are important to anyone who studies or practices chemistry. Chapter 1 discusses the role of writing in chemistry, and Chapter 2 treats the responsibilities that the chemist as a writer has to the scientific community and the public.

The next two chapters cover some of the basics required for taking a chemistry course, the survival skills that are especially useful for beginning chemistry students. Chapter 3 treats necessary study skills, including how to get the most out of a chemistry lecture and how to read a chemistry textbook. Chapter 4 discusses the chemistry laboratory, with a particular emphasis on how to write the laboratory report.

The following three chapters are devoted to the types of reading and writing that are characteristic of the further study of chemistry. Chapter 5 explains how to read and interpret scientific articles. Chapter 6 discusses how to do searches of the scientific literature and how to write literature reviews. Chapter 7 offers an introduction to preparing research proposals.

The last chapter will be of interest to all students of chemistry. Chapter 8 is about chemistry and the public. This chapter is concerned with how the public views chemistry and how a knowledge of chemistry can be used to inform the public and to influence public opinion. Some of the basics of science writing, a useful skill for technically trained people, are explained in this chapter.

Part of the strategy of this book is to provide examples of writing about chemistry. As you read these selections, we hope that you come to see chemistry as a human experience and an organized attempt to understand the nature of the physical universe. Because writing about a subject is the most effective way for most people to develop their thinking, writing assignments, mostly quite short, have also been included within the text and at the ends of the chapters.

ACKNOWLEDGMENTS

This book grew out of the Writing Across the Curriculum Project at Worcester Polytechnic Institute for which we are grateful to the General Electric Foundation for support. We wish to thank our editor at HarperCollins, Patricia Rossi. Martha Catherine Trimbur provided valuable help in preparing the index. H. B. wishes to thank his wife Barbara for advice and encouragement.

We would also like to thank the many people who reviewed this manuscript. Their comments and suggestions were greatly appreciated:

Frank Andrews
University of California–Santa Cruz

Robert Desiderato
University of North Texas

Thomas M. Dunn
University of Michigan

Phillip S. Lamprey
University of Massachusetts–Lowell

Harold Spevack
Borough of Manhattan Community College of CUNY

John Suchocki
Leeward Community College

Chris Thaiss
George Mason University

Lili Fox Vēlez
Carnegie Mellon University

John A. Weyh
Western Washington University

Herbert Beall
John Trimbur

A Short Guide to Writing about Chemistry

1

Writing and Chemistry

Learning chemistry involves more than learning chemical facts and concepts. It also involves learning how to read and write as chemists do.

To learn chemistry, you will need to study the material presented in your textbook, in lectures, and in the laboratory. There will be definitions to remember, concepts to acquire, problems to solve, and procedures to learn. And there will no doubt be quizzes, examinations, and problem sets to test and measure how well you have learned the material.

But chemistry is not just a body of information for students to learn and be tested on. Chemistry also refers to the work chemists do—how they define problems, design experiments, and interpret results. From this perspective, to learn chemistry means learning how chemists think and act. It means learning how chemists go about their daily work, what their habits of thought are, and how they communicate with each other. Whether or not you are planning to become a chemist, understanding the patterns of thinking, problem-solving strategies, and ways of reading and writing that define the study and practice of chemistry can be valuable.

Chemistry is a way of asking questions about the physical world, of figuring out ways to answer these questions experimentally, and of communicating the significance of the results to other people. To think of chemistry as an active, organized process of investigating the physical world and communicating the results can help you understand the purposes of chemical knowledge and research.

Whatever your major is, learning how chemists formulate problems and communicate their work can help you read and evaluate the results and claims of research in chemistry. Currently, there are increasing numbers of public policy issues that require knowledge of chemistry. What is the effect of acid rain on the environment? Should the government regulate the use of dioxins? What is a toxic substance? Is lead a serious public health problem? Chemistry is a practical tool that can help answer these questions. As a citizen, you need to understand how chemistry works in order to make informed and responsible judgments.

A chemistry class is a good place to learn how to write the clear, logical, and concise prose that chemists—and other professionals—value. It may well be the case that most of the writing you have done in school has been in English, history, and other humanities and social science courses. For this reason, students sometimes complain when they are assigned writing in chemistry classes. "This isn't an English course," they say, implying that it is somehow unfair to require good, effective writing in science, mathematics, or engineering courses. But if part of learning chemistry is learning what chemists do, then reading and writing make a lot of sense in a chemistry class because chemists devote a good deal of their time to these activities. In fact, as you will see in the next section, reading and writing are key ways that individual chemists participate in the ongoing work of the larger community of chemists.

Finally, a good reason for writing in chemistry, whether or not you are going to become a chemist, is that most professional fields—such as law, business, government, medicine, and so on—value highly an individual's ability to communicate complicated technical and scientific information. Surveys to determine what employers are looking for in prospective job applicants uniformly rank communication skills at or near the top. Learning to read and write chemistry can be good practice to acquire such skills and to enhance your ability to secure satisfying employment when you graduate.

Like other academic or professional fields, chemistry is not simply a body of knowledge. It is also the community of practicing chemists who need to communicate with each other. Throughout this book, you will be introduced to some of the forms of writing chemists rely on to make this communication possible, as well as the style of writing they use.

WHAT DO CHEMISTS READ AND WRITE ABOUT?

Let's take a look at the connections among reading, writing, and the actual practice of chemistry. Whether they work in industry or in academic science, chemists typically spend about half their time reading and writing. Most chemists read a number of scientific journals regularly to keep up with ongoing work in their field. They do this to stay current with recent developments, to know what other researchers in their particular field are doing, to shape the direction of their own research, and to be active participants in the community of chemists. Chemists in colleges and universities who teach introductory and advanced chemistry courses also need to know about the current state of knowledge so that they can incorporate recent findings and ideas into their lectures and class discussions.

To put it another way, chemists read in order to write. Many chemists, for example, read in order to write proposals for grants to acquire the funding or support necessary to pursue their research interests and to publish the results of their work. A research proposal is one of the most important forms of writing chemists do. Grant funding is crucial to scientific researchers in colleges, universities, and research institutes, to provide them with the money to buy materials and supplies, to hire technical help, and to pay part of their own salary. In many colleges and universities, a faculty member in science is paid only a part of his or her salary by the institution and must raise the rest from external funding agencies such as the National Science Foundation or the National Institutes of Health. By the same token, chemists in industry write research proposals to acquire the support they need to solve problems and develop new products.

To be successful, a research proposal must present an effective argument that will persuade the panel of experts who read and evaluate proposals for grant-funding agencies or the supervisors in industry that the research proposed is significant, that the methods of pursuing this research are well designed, and that the researcher is in fact capable of carrying out the research. If an individual chemist has already published articles in his or her field of research, his or her grant application, of course, will be all the more persuasive.

There is an interesting cycle of writing at work here. Chemists need to write effective research proposals in order to do research so that they can publish scientific articles and develop new products and industrial processes. At the same time, the articles chemists publish will help them to secure future funding for their research so that they can continue to publish or to develop products.

The scientific articles chemists write are similar in many respects to the laboratory reports you may already have written in chemistry courses or will be writing in this one. Like a good laboratory report, a well-written article in chemistry will define a significant problem, explain how the work reported relates to previous research in the field, describe the materials and methods used in the research, present the results, and then interpret these results in the context of other work and basic chemical principles. The main difference between a scientific article and a laboratory report is that the research used for a scientific article takes place over a longer period of time and therefore produces more data than is possible in one laboratory session.

Once a chemist prepares a manuscript and sends it to the editor of a journal, there is still a good deal of reading and writing to be done. The editor sends out the manuscript to be reviewed by two or three other chemists in the field. They will read the manuscript, write comments, and recommend whether the editor should publish the article, reject it, or ask the author to revise and resubmit. Typically, the editor will neither accept the article nor reject it outright. It's more likely that the author will be asked to revise the article, perhaps to perform additional experiments, and then resubmit the manuscript for another review. At this point, the editor may accept the article in its revised form or ask for further revision before publication.

If this sounds like a lot of reading and writing, we can add to it other kinds of writing chemists do in the course of their daily work. They write progress reports on their research for their supervisors or for the agency that funded it. They write reports and memoranda for their company or for the chemistry department of the college where they work. In addition, academic chemists write and revise their lectures periodically so that they tell an orderly and coherent story about the state of chemical knowledge. A good lecture—and the preparation of slides, handouts, and other course materials—is itself a major writing task, equivalent to a good-sized paper, even if the lecturer writes notes instead of a complete text of the lecture. Chemistry faculty also write letters of recommendation for students who are applying to graduate and medical schools. Chemists may also write informative brochures and leaflets and letters to politicians and newspapers on behalf of environmental causes or other public interest concerns.

And we haven't even mentioned the reading and writing that chemists do daily in their laboratories. This reading and writing is so intimately and inextricably tied to the everyday practice of chemical research that chemists may not even think they are reading and writing. But, as all scientists do, chemists keep a laboratory notebook, in which they write daily notes about the experiments they are performing and the data that results. As you will see in Chapter 4 "Writing Laboratory Reports," the chemist's notebook is the primary document of a chemist's work.

There are also other kinds of writing chemists do in the lab. For example, they are constantly marking test tubes, beakers, and flasks to keep an accurate account of the materials and the amounts they are using. They are also constantly writing down the figures they read off the various meters and scales of the scientific equipment that makes up their lab. Chemistry, in other words, like the other sciences, is a practice of constantly recording what a scientist does and sees. Writing is not something that happens once an experiment is finished. Writing and reading are integral parts of experimental work.

To see a chemist's work from the inside is to encounter how thoroughly reading and writing pervade a chemist's professional life and laboratory work. As the sociologists of science, Steve Woolgar and Bruno Latour, put it, scientists are "compulsive and almost manic writers" who "spend the greatest part of their day coding, marking, altering, correcting, reading, and writing."

HOW WRITING ABOUT CHEMISTRY CAN HELP YOU BECOME A BETTER WRITER

Learning to read and write as chemists do can help you learn chemistry from the inside, as well as help you become a better writer.

Learning to write means learning how to write knowledgeably and competently in a particular field. Every profession has its own particular writing practices adapted to its own goals and purposes. For this reason, one of the best ways to improve your writing is to learn the purposes and forms that writers in a particular community use to communicate to other members of the community. By immersing yourself in the kind of writing that takes place in a particular community, you can become more aware of the factors that define the role of writing in that field.

Once you have a working knowledge of the basic mechanics of handwriting, spelling, and grammar, then learning to write is really a matter of learning how and why writers in particular fields write as they do. What kind of writing is valued in particular academic and professional communities? What functions does writing perform?

How do writers know what tone their writing should take? How do they know what other members of their community will be interested in reading? How do writers position themselves in relation to their material, to their readers, and to what others have already written about a topic? To find answers to these questions, it will help to analyze the writer's situation in his or her academic or professional field.

Let's begin by looking at four components that make up the writer's situation in the community of chemists:

- the writer,
- the purpose of writing,
- the forms of writing, and
- the audience.

We'll take a look at each component separately, but, as you will see, they are interrelated. Putting them all together can shed light on how chemists write.

Who Is the Writer?

At first glance, this question may sound silly. Isn't the writer the person who does the writing? In one sense, of course, this is true. But there is more to it. When chemists write research proposals or articles, they are not simply writers. They write as practicing chemists. In fact, they write *because* they are chemists, and in chemistry, a writer is not just anyone but rather is a person—a researcher, a teacher, an environmentalist—who asks questions of chemical phenomena in a disciplined way. Learning to write in chemistry amounts to learning what questions the community of chemists considers meaningful, what evidence they take to be legitimate, and what interpretations they find persuasive.

Writing in chemistry, in other words, means taking a particular stance toward the physical universe, the materials you are writing about, and the language of chemistry. This stance comes across in a particular voice that characterizes the writing of chemists—a tone that is *precise*, *concise*, and *objective*. It is *precise* in that it uses accurately language, terminology, systems of measurement, and so on that have been agreed on by the wider community of chemists. It is *con-*

cise in that writers limit themselves to what is relevant to the scientific work being described and discussed. It is *objective* in that the focus of the writing is on the chemical phenomenon being investigated, not on the perceptions, beliefs, or opinions of the writer.

What Is the Purpose of Writing in Chemistry?

Chemists' purposes in writing are closely tied to who the writer is in chemistry. In chemistry, the basic purposes of writing are to clarify the meaning of your work, to inform your audience about its results (whether the work consists of reading a textbook or performing laboratory experiments), and to offer a persuasive interpretation of the results, based on the available evidence and a reasoned argument that sticks to the data. To do this, chemists typically define what are considered significant problems, design research to explore these problems, and then explain how the results shed light on the problem under consideration. Chemists seek to inform and persuade their readers within the limits of well-defined problems.

What Forms Do Chemists Use?

The specialized forms of writing that communication in chemistry depends on are linked to the chemist's stance toward the physical universe and the purposes of writing in chemistry. The standard forms of writing in chemistry are really conventions to help chemists communicate their ideas and the results of their work. Standardizing the forms actually makes it easier and more convenient to communicate because all the participants—the writers and readers—in the field of chemistry share a common means of communicating with each other. To learn to read and write in these conventionalized forms, therefore, is a way of gaining access to the collective work of the wider community of chemists. They are, in short, important tools of the trade all chemists learn to use, not unlike laboratory equipment or materials and supplies.

These standard forms include,

- Laboratory notebooks,
- Laboratory reports,
- Summaries of published articles,

- Literature reviews,
- Research proposals, and
- Research articles.

Who Is the Audience?

By now it may be evident that the audience for writing about chemistry mainly consists of chemists. The problem, though, is that in chemistry courses students know that their audience is not really the community of chemists but their teacher, who is going to read, evaluate, and grade their writing. There is no point in denying this fact, but it is important at the same time to recognize the potentially distorting effects this relation can have on student writing in chemistry. It puts students in the position of writing to perform for an evaluator instead of writing to communicate to colleagues in the field, to inform and persuade them of the significance of the writer's work. One way to deal with this problem is to imagine that your audience consists of other students in your chemistry class. Write to them, explaining as clearly as you can the chemical concepts and facts that you are learning.

Exercises

1. The following selection is from a standard work in the field of chemistry, *Active Carbon* by Bansal, Donnet and Stoecki. In this passage, the authors define activated carbons and their uses. In many respects, this passage can be taken as a model of *precision* in writing about chemistry. As you read this selection, notice what makes the writing precise. Point to particular words, phrases, or sections to explain why the writing is precise. Compare your reading of the selection to that of others in your class.

    The term *activated carbon* in its broadest sense includes a wide range of amorphous carbon-based materials prepared to exhibit a high degree of porosity and an extended interparticulate surface area. These are obtained by combustion, partial combustion, and thermal decomposition of various carbonaceous substances. These materials may be granular or in powdered form. The granular form is characterized by a large internal surface and small pores, whereas the finely divided powdered form is associated with larger pore diameters but a smaller internal surface.

Activated carbons are excellent adsorbents and thus are used to purify, decolorize, deodorize, dechlorinate, detoxicate, filter, or remove or modify the salts, separate, and concentrate in order to permit recovery; they are also used as catalysts and catalyst supports. These applications of active carbons are of interest to most economic sectors and concern areas as diverse as the food, pharmaceutical, chemical, petroleum, mining, nuclear, automobile, and vacuum industries as well as the treatment of drinking water, industrial and urban wastewater, and air and gas. Nearly 80% (220,000 tons/yr) of the total active carbon is consumed for liquid phase applications where both the granulated and powdered forms of active carbon are used. The use of powdered carbon is more ancient and generally involves processing of food and drinking water. The total consumption of active carbon in gas phase applications is around 60,000 tons/yr, which includes exclusively the granular form of the active carbon, which may be extruded or pounded, the principal uses being in the purification of air, recovery of gold, and cigarette filters. The consumption of active carbon is high in the United States and Japan, which together consume two to four times more active carbon than the Western European countries. The per capita consumption of active carbon per year is 0.5 kg in Japan, 0.4 kg in the United States, 0.2 kg in Europe, and 0.003 kg in the rest of the world.

2. The two following passages were written by students in response to the assignment, "Explain why the law of definite proportions or constant composition of compounds leads to Dalton's atomic theory of matter." Compare them in terms of *conciseness*. Is one more successful in emulating the concise tone associated with chemical writing? Remember that being concise does not necessarily mean being brief but rather means being relevant to the scientific task. Be specific in identifying irrelevant passages.

   a. The law of definite proportions says that a compound always has the same proportions by weight of each of the elements that it is made up of. That is, the composition of a compound is always the same no matter where it comes from or how it is made. Common salt has the same percentage (by weight) of sodium and chlorine no matter whether it comes out of a Morton salt box or is crystallized from water from the Indian Ocean. Actually, if there are impurities in the salt, like another compound added to keep it from clumping up on a humid day, the composition will vary a little but the composition of the actual sodium chloride will always be the same. Now how do we explain this in terms of what is going on at the level we can't see? Well, if the sodium and chlorine could exist in pieces that were unlimited in how small they were, that is could be infinitely small, then you couldn't say how much sodium could combine with how much chlorine. Therefore, the units of sodium and chlorine must have some fixed size even if this size is pretty small. That's the atomic theory.

b. The idea that the composition of a compound is always the same means that the elements that make it up must be comprised of units of finite size. If the elements were infinitely divisible, any combination of elements would be possible. The finite sized and indivisible unit is Dalton's atom. Furthermore, the constant composition of the compound means that the atoms that comprise it must always combine in precisely the same way. Therefore, according to Dalton's theory, all atoms of the same element should be the same.

3. The following passages were written by students who were asked to pose a question based on something they have ob-

served in the world that the field of chemistry might address. Is one of the passages more *objective* in tone and stance than the other? Consider whether each adequately distinguishes between fact and interpretation. Does either one use unstated premises?

a. `I want to understand how industrial pollution has caused the sky at sunrise and sunset to appear orange and red and, less frequently, light green. I would like to identify the pollutants that cause these effects by demonstrating them in the laboratory.`

b. `I am interested in understanding the colored effects that occur in the direction of the sun at sunrise and sunset. The colors I have observed have included the range from orange to red and, less frequently, light green. I would like to investigate the molecules and conditions that cause these effects and whether similar effects can be demonstrated in the laboratory.`

Writing Assignments

1. Interview a chemistry major, graduate student, or professor about the kind of work he or she does in a chemistry lab. Find out what problems or questions your subject is working on and what he or she hopes to learn. Ask the person you're interviewing how much and what kind of writing he or she does in the course of work. What are the person's feelings about writing? How did the person learn to write in chemistry? What tips can the person offer you? Write a short profile of the person as a chemist and a writer.

2. Read the preface and introduction to the chemistry textbook you are using in your course. Write a short summary of how the preface and introduction present chemistry to students and analyze the image of chemistry they offer.

2

Writing and Scientific Responsibility

Like any professional field, science has a code of ethics that practitioners are obliged to follow. Unlike other fields, however, this code is not written down. There is nothing comparable, say, to the Hippocratic Oath in medicine. Instead, the ethics that govern the practice of science are based on an unstated but nonetheless clearly understood set of values that scientists share and expect each other to uphold.

In this chapter, we will look at the ethics of science and how they relate to the work of scientific writing.

SCIENTIFIC HONESTY

The foundation of scientific ethics can be summed up in one word—honesty. The integrity of science as a disciplined investigation of the physical and natural world depends on the honesty of everyone involved in scientific work. Honesty is critical because the practice of science is really the collective activity of a community of investigators. The ability of a scientific field to make progress in understanding some aspect of the physical universe or in solving a particular problem typically relies on the work of many scientists in a number of laboratories. It is rare that an individual scientist working in isolation makes a breakthrough that clarifies a concept or solves a problem. Instead, the contributions of many scientists form a kind of ongoing conversation in which the participating scientists report research, make claims, ask questions, challenge, criticize, correct each other, and modify their own views and research strategies. For this conversation to proceed, the participants need to trust each other. To judge and comment on each other's work, scientists must be able to assume that the data and interpretation in a scientific paper represent the author's own work reported honestly without fabrication.

Plagiarism, which occurs when an individual presents someone else's work as his or her own, and falsified data, therefore, represent serious breaches of scientific ethics. Such dishonesty threatens to break the bonds of trust in the collective conversation that enables scientists to produce reliable knowledge in science.

By the same token, scientists have to be honest about the contributions made by co-workers to a particular piece of research. If you look through some recent scientific journals that publish experimental articles, you will notice that most of the articles list a number of authors, sometimes three or four but in other cases many more. Part of scientific honesty involves acknowledging the collective work that goes into research.

Another reason that honesty is critical to the practice of science has to do with the fact that scientists are responsible to those in related fields, such as medicine, engineering, and industry, who apply the results of scientific research in practical ways. A doctor, for example, has to be confident that a new drug developed in a laboratory has the properties and effects researchers describe before prescribing it to patients. Similarly, an engineer has to be confident that the yield point of a particular substance is what a materials scientist says it is before using it in construction or in industrial processes. Scientists are also responsible to the funding agencies that support their research. Because so much scientific research is funded by the federal government, scientists are ultimately responsible to taxpayers and the public.

In advanced industrial nations such as the United States, science and technology are pervasive facts of life, and for this reason, the credibility of scientific work is crucial to the basic functioning of society. In many respects, the health, safety, and prosperity of a people depend on the honesty of scientific research. The public as well as local, state, and federal governments rely on the honesty of scientific researchers in many ways—to set proper standards for water purity, to develop occupational safety guidelines in workplaces, to set policies about the disposal of toxic wastes, to make decisions about the use of nuclear energy. Accordingly, scientific fraud or dishonesty threatens not only the professional integrity of science but society as a whole.

Nonetheless, as you may be aware from newspaper and television accounts, there are instances of scientific fraud. It would be dishonest of us not to face these occurrences in a straightforward manner. To understand science as it is really practiced requires a frank look at this feature of scientific life.

Scientists are human, and despite the code of ethics to which they subscribe, researchers can make errors of ethical judgment. The following section considers how and why the principle of scientific honesty is sometimes violated.

UNDERSTANDING THE COMPETITIVE WORLD OF SCIENCE

We have described the practice of scientific work as a collective effort, the results of which depend on the cooperation of a community of scientists. At the same time, however, it is important to note that while progress in science depends on cooperation, scientific work may often be intensely competitive. Because scientists tend to work along parallel lines of investigation and consider similar problems, individual scientists and laboratories sometimes vie with each other to be the first to publish significant findings. In science, the success and prestige ascribed to particular scientists and laboratories are based on the priority of discovery and publication. In many ways, the worst fear of a scientist is that he or she will be "scooped" by a rival laboratory. Scientific research is often a race to get there first—whether the race is to be the first to figure out the structure of DNA, which formed one of the central dramas in science in the early 1950s, or more recently to identify the gene responsible for hereditary breast cancer.

There are potential tensions between cooperation and competition at the center of scientific work. On the one hand, the participants in a scientific field jointly create an intellectual context that benefits everyone. On the other hand, despite these cooperative efforts, scientists are aware that individual careers, prestige, and grant funding are determined competitively. Usually these two pressures—the collaborative and the competitive—work together to move both the field and individual investigations forward. At other times, however, these pressures can cause a researcher to take ethical shortcuts such as not giving proper credit to a co-worker or falsifying data.

Scientific dishonesty must be seen in the context of these pressures and in the context of the current moral climate in the United States. At a time in our history when political scandals in government, insider-trading violations on Wall Street, and increasing levels of cheating in high school and college are front-page news, cases of scientific fraud are part of a more general and disturbing dishonesty in our culture. This does not excuse those scientists who commit fraud but it may help to explain their actions.

The point finally, though, is to use instances of fraud as an occasion to raise basic issues about the ethics and responsibilities of professionals. As you consider your own actions as a student on your way to a career, you need to take these issues into account—to think about the kind of society and ethics your actions will bring into being and what kind of world you want to live in.

The following excerpt from *The Double Helix*, Nobel Laureate James D. Watson's book about the discovery by Watson and Francis Crick of the structure of DNA, offers an unusually candid picture of the world of science from the inside. Written for general readers, *The Double Helix* is a popular account of Watson and Crick's race to be the first to identify the structure of DNA. In this passage, Watson describes his and Crick's reaction to the wrong direction taken by Linus Pauling, one of their chief competitors. Pauling himself was an extremely charismatic and influential figure and one of the giants of twentieth-century chemistry.

The Double Helix

James D. Watson

At once I felt something was not right. I could not pinpoint the mistake, however, until I looked at the illustrations for several minutes. Then I realized that the phosphate groups in Linus' model were not ionized, but that each group contained a bound hydrogen atom and so had no net charge. Pauling's nucleic acid in a sense was not an acid at all. Moreover, the uncharged phosphate groups were not incidental features. The hydrogens were part of the hydrogen bonds that held together the three intertwined chains. Without the hydrogen atoms, the chains would immediately fly apart and the structure vanish.

Everything I knew about nucleic-acid chemistry indicated that phosphate groups never contained bound hydrogen atoms. No one had ever questioned that DNA was a moderately strong acid. Thus under physiological conditions, there would always be positively charged ions like sodium or magnesium lying nearby to neutralize the negatively charged phosphate groups. All our speculations about whether divalent ions held the chains together would have made no sense if there were hydrogen atoms firmly bound to the phosphates. Yet somehow Linus, unquestionably the world's most astute chemist, had come to the opposite conclusion.

When Francis was amazed equally by Pauling's unorthodox chemistry, I began to breathe slower. By then I knew we were still in the game. Neither of us, however, had the slightest clue to the steps that had led Linus to his blunder. If a student had made a similar mistake, he would be thought unfit to benefit from Cal Tech's chemistry faculty. Thus, we could not but initially worry whether Linus' model followed from a revolutionary re-evaluation of the acid-base properties of very large molecules. The tone of the manuscript, however, argued against any such advance in chemical theory. No reason existed to keep secret a first-rate theoretical breakthrough. Rather, if that had occurred Linus would have written two papers, the first describing his new theory, the second showing how it was used to solve the DNA structure.

The blooper was too unbelievable to keep secret for more than a few minutes. I dashed over to Roy Markham's lab to spurt out the news and to receive further reassurance that Linus' chemistry was screwy. Markham predictably expressed pleasure that a giant had forgotten elementary college chemistry. He then could not refrain from revealing how one of Cambridge's great men had on occasion also forgotten his chemistry. Next I hopped over to the organic chemists, where again I heard the soothing words that DNA was an acid.

By teatime I was back in the Cavendish, where Francis was explaining to John and Max that no further time must be lost on this side of the Atlantic. When his mistake became known, Linus would not stop until he had captured the right structure. Now our immediate hope was that his chemical colleagues would be more than ever awed by his intellect and not probe the details of his model. But since the manuscript had already been dispatched to the *Proceedings of the National Academy,* by mid-March at the latest Linus' paper would be spread around the world. Then it would be only a matter of days before the error would be discovered. We had anywhere up to six weeks before Linus again was in full-time pursuit of DNA.

Though Maurice had to be warned, we did not immediately ring him. The pace of Francis' words might cause Maurice to find a reason for terminating the conversation before all the implications of Pauling's folly could be hammered home. Since in several days I was to go up to London to see Bill Hayes, the sensible course was to bring the manuscript with me for Maurice's and Rosy's inspection.

Then, as the stimulation of the last several hours had made further work that day impossible, Francis and I went over to the Eagle. The moment its doors opened for the evening we were there to drink a toast to the Pauling failure. Instead of sherry, I let Francis buy me a whiskey. Though the odds still appeared against us, Linus had not yet won his Nobel.

Writing Assignments

1. Describe your response to Watson's description of his discovery of Pauling's error. Does Watson's description change or confirm your image of how science is practiced? Write a short essay that explains your response.
2. In this episode, Watson focuses largely on the competition between his work and Pauling's work on the structure of DNA. Is there a sense in which this rivalry could be thought of as cooperative as well as competitive? Write a short essay that explains your answer.
3. Do you think that Watson's personal qualities as revealed in this excerpt detract from his stature as a scientist? Write a short essay that explains your opinion.
4. In the next-to-last paragraph of this selection, Watson mentions bringing Pauling's manuscript "for Maurice's and Rosy's inspection." "Rosy" here refers to Rosalind Franklin, one of Watson and Crick's co-workers. There is some question as to whether Franklin has received proper credit for her contributions to Watson and Crick's work. Do some research to answer this question and write a short essay that develops your view of whether Franklin did or did not receive proper credit.

A CASE OF POSSIBLE DISHONESTY

A notable case of possible scientific dishonesty involved a paper published in the prestigious journal *Cell* in 1986 by Noble Laureate David Baltimore, Thereza Imanishi-Kari, and four co-workers. Margot O'Toole, a postdoctoral fellow working in Imanishi-Kari's lab at the time, charged that experimental data crucial to the central claim in the paper were fabricated. According to O'Toole, the experiments described had in fact never been performed.

O'Toole's charge of scientific fraud led to a series of investigations, first by Tufts University and Massachusetts Institute of Technology, which found no evidence of misconduct, and later by the National Institutes of Health's Office of Scientific Integrity, which upheld O'Toole's allegations. Ultimately, the case worked its way into the halls of Congress, when Rep. John D. Dingell convened his Subcommittee on Oversight and Investigation to look at the accusations O'Toole had made and how they had been handled by the various investigative bodies.

This case is notable in part because of the standing of David Baltimore, one of the leading scientific researchers in the nation. Many scientists came to Baltimore's support and condemned Dingell's hearings as unnecessary and inappropriate interference by the federal government in the affairs of science. Others argued that the case shows the inability or unwillingness of major scientific research institutions to properly investigate charges of misconduct. Still others were concerned with the fate of "whistleblowers" such as O'Toole and the pressures within the scientific community not to report incidents of fraud or misconduct. O'Toole was fired from her job by Imanishi-Kari and it was four years before she was able to find another research position.

Below we reprint a number of documents concerning this case of possible dishonesty. We present first David Baltimore's view of the controversy, "Baltimore's Travels," written in 1989; next a report on the investigations, "Scientific Whistleblower Vindicated," written in 1991; and finally the retraction of the paper in 1991.

Baltimore's Travels

David Baltimore

For the past three years, I have been at the focal point of an investigation into the accuracy of a scientific paper (published April 1986 in the journal *Cell*) of which I was a coauthor. The investigation, which wound up before a subcommittee of the U.S. House of Representatives and was reported on the front pages of newspapers, continues to this day. My purpose in writing this article is to explain how a routine scientific dispute came to be adjudicated in such an unusual setting, how we chose to defend ourselves, and the scientific and political issues involved. One central issue was apparent to me from the beginning, and it remains central today: Who should judge science? Along the way I came to identify several other issues:

- What is the worth of scientific collaborations, and what are the duties and responsibilities of those involved in them?
- Is error in science inherently bad and should it always be corrected where found?
- How does one distinguish between error and fraud?
- And does science do an adequate job of policing itself?

The Background

In the early 1980s, I began a series of experiments to explore the production and control of antibodies—proteins made by immune cells that participate in the recognition and destruction of invading microorganisms. The experiments employed an exciting new technology, the insertion of new genes into mouse embryos to produce "transgenic mice." The work required the collaboration of several laboratories with expertise in different areas of molecular biology and immunology.

Out initial goal was to produce transgenic mice carrying a new antibody gene. The experiments worked quite well and were described in two papers that predated the *Cell* paper involved in the current controversy. Studies of the expression of the inserted "transgene" provided new insights into the regulation of antibody production. A next step in this work was to examine how the new antibody gene affected the activity of natural antibody genes. For these experiments, I approached Professor Thereza Imanishi-Kari at the MIT Cancer Center. We designed a two-pronged study: Her laboratory would analyze the antibody proteins produced by the transgenic mice, and my laboratory would characterize the active antibody genes. One purpose of this parallel design was to provide an internal mechanism for cross-checking results. For example, if the protein analysis—always a tricky and difficult task—should give inconsistent results, the genetic analysis might reveal the source of the problem.

One key observation was that the inserted antibody gene had a far greater influence than previously expected on antibody production by the resident mouse genes. Much of our analysis focused on the region of the antibody molecule that enables each antibody to recognize a specific foreign structure; scientists refer to this region as having an "idiotype." Our previous studies had shown that many immune cells from the transgenic mice were secreting antibodies with idiotypic regions matching those of the new gene. We had assumed that these antibodies were products of the transgene, but the new experiments told a different story. They showed that most of the relevant antibodies were derived from normal antibody genes inside the mouse cells. Through some unknown mechanism, the presence of the inserted gene had induced the cells to adopt a particular idiotypic pattern.

The *Cell* paper describes two possible explanations for this mimicry. It is important to recognize, however, that the paper and the experiments it describes are just one link in the chain of scientific evidence. We fully expected that other laboratories would extend the experiments into new directions and publish their own interpretations. Each new effort would represent a small step forward in understanding how the immune system responds to external challenges. Future studies might offer new insights into derangements of the immune system that cause autoimmune diseases such as arthritis, lupus, and certain life-threatening anemias.

The Paper is Challenged

At the outset, the substance of the dispute was not unlike others that occur regularly in biology labs. It was simply a disagreement over scientific matters between two scientists: Professor Thereza Imanishi-Kari, now at Tufts University, a leading immunologist who specializes in the analysis of antibodies; and Dr. Margot O'Toole, a postdoctoral researcher on a one-year appointment in the Imanishi-Kari laboratory, who was not a participant in the study but worked on a related question.

Dr. O'Toole had reviewed the draft of our paper, made a number of helpful suggestions, and indicated no concerns prior to publication. Shortly after the paper was published, however, O'Toole expressed reservations in a most unusual fashion: Rather than discuss her objections with the authors (the usual procedure), she complained to an outside authority—not MIT, which employed both O'Toole and Imanishi-Kari, but Tufts University, which was considering whether to appoint Imanishi-Kari to its faculty. The gist of her complaint then—it has changed from time to time over the intervening three years—was that the conclusions of the paper were not supported by the scientific evidence it presented.

Tufts appointed a panel to review O'Toole's charges about the paper and also appointed a special committee to review Imanishi-Kari's qualifications and reconsider her appointment. The scientific review panel, headed by Tufts professor Henry Wortis, found that O'Toole's complaints involved matters of interpretation and, because the paper's central thrust was unaffected, did not recommend any corrective action. Following its two reviews, Tufts appointed Imanishi-Kari to a tenure-track position.

O'Toole then took her complaints to MIT, which appointed Professor Herman Eisen, an eminent immunologist, to review them. He found that she had made an appropriate expression of scientific concern, but the errors he observed in the paper were not of sufficient importance to require correction. Further, he concurred with the Wortis panel's decision that although O'Toole's interpretation might differ from that of her supervisor, the paper's conclusions were sound.

In these reviews, completed by early summer of 1986, all the issues were scientific. No one had accused anyone of unethical or criminal behavior; O'Toole simply said she thought the conclusions of the paper were not borne out by the data generated in the Imanishi-Kari laboratory. After the second review, I thought that the matter was closed.

Meanwhile, Charles Maplethorpe, who received his Ph.D. for work done with Dr. Imanishi-Kari but who had left her laboratory before the paper was published and had not been involved in the study, got himself involved. Maplethorpe contacted two scientists at the National Institutes of Health (NIH), Walter Stewart and Ned Feder, who had made reputations for themselves by publishing papers analyzing cases of previously demonstrated fraud in science. They received from O'Toole copies of 17 pages of laboratory notes taken from the Imanishi-Kari laboratory. These pages, of more than a thousand pages collected during the study, included data from a number of failed experiments. On the basis of the 17 pages, plus conversations with O'Toole and Maplethorpe, Stewart and Feder mounted a challenge of their own.

The Stewart and Feder challenge soon developed into a *cause célèbre* because of the manner in which they conducted it. First they wrote a lengthy manuscript clearly charging that our paper was consciously misleading. The Stewart-Feder manuscript was submitted to a number of journals, all of which rejected it. Frustrated by their inability to publish what journal editors told them was not a scientific article that could be refereed, Stewart and Feder went public. They circulated the manuscript widely to scientists, asking for comment. They also began speaking about their "investigation" on university campuses and at scientific meetings and offered to send a complete file of correspondence to anyone asking for it.

That file contained a large number of letters and memoranda they had exchanged with me, my fellow authors, other scientists, editors, and NIH officials. It was a highly selective file, containing, for example, little or nothing of the correspondence that by this time they were having with Margot O'Toole and with staff members of congressional committees. In much of this correspondence, and in the speeches of Stewart, my name and those of my colleagues were frequently juxtaposed with the names of people who have been found guilty of fraud. In other words, Stewart and Feder utilized a classic propaganda technique—guilt by association. Because of their hostile acts, I asked the NIH to conduct a formal review to determine the accuracy of the *Cell* paper.

This activity by Stewart and Feder continued throughout 1987 and early 1988, at which time I learned from a newspaper reporter that two congressional committees investigating fraud and misconduct would shortly hear from Stewart and Feder about our paper. One committee was the Oversight and Investigations Subcommittee (O&I) of the House Energy and Commerce Subcommittee, chaired by Congressman John Dingell (D-Mich.); the other was the Human Resources and Intergovernmental Relations Committee of the House Government Operations Committee, chaired by Congressman Ted Weiss (D-N.Y.).

We were not notified of these hearings nor were we permitted to answer the charges against us. The Weiss subcommittee hearing referred only obliquely to our dispute with Stewart and Feder and did not mention by name the authors of the paper. At the O&I hearing, however, my name was used liberally and the words "fraud," "mis-conduct," and "misrepresentation" were casually bandied about. Although I was not personally accused of having committed fraud, the implication was clearly made at the hearing that at least one of the authors had.

Meanwhile, NIH (which funded the research) was convening a panel, consistent with my earlier request, to review the dispute. The panel's start was delayed by inappropriate appointments, however—two of its initial members had coauthored papers with me—and was slow getting under way.

Defending Against Attack

In the wake of the congressional hearings of April 1988, it became clear that my own reputation and those of my coauthors were being sullied by continuing acrimonious attacks from Stewart and Feder, who had allied themselves with Peter Stockton, a policy analyst on the staff of the O&I subcommittee. Soon after, Stewart and Feder were detailed from NIH to work with Stockton. Clearly, the authors needed to mount a defense.

For a time we got by with just my own staff, consisting of the administrator in my office and a part-time public relations person who does work for the Whitehead Institute, plus assistance from my personal lawyer. Before long it became apparent that we needed help from people familiar with the procedures of Congress and the workings of Washington politics. We retained Washington counsel and spoke with others knowledgeable in the ways of that city.

We learned something about the nature of such congressional investigations: The effort to resolve the dispute would not be primarily a scientific or a legal one. Rather, it would be largely directed at press coverage. We would not have a "day in court" in the sense of being informed in advance of any charges or complaints about our work. We would not have the opportunity to confront and question our accusers, to present our own evidence or witnesses, or have any assurance that a record could be developed and a decision based on it. There would likely be no final ruling from a congressional panel; instead, we would be judged in the "court of public opinion."

I was advised that although the alleged errors (or fraud or misconduct, as they came to be referred to) did not take place in my laboratory, I was nevertheless the real target of Stewart, Feder, and Stockton because of my visibility. Thus I agreed to lead the defense. One of my first defensive actions was to write a letter to about 400 of my scientific colleagues, acquainting them with my side of this matter. This action was reported in some newspapers.

Meanwhile, the third scientific review of the *Cell* paper was in progress. The NIH Review Panel deliberated through the last half of 1988. Like the two reviews before it, the NIH panel reported that there were errors but the findings of the paper were sound. Fraud, misconduct, or misrepresentation were explicitly rejected. Unlike the other two reviews, this one suggested corrections in the journal. Although we disagreed with the panelists—as did *Cell*'s editor—corrections about the errors were submitted and published.

Meanwhile, unbeknownst to us, Stewart, Feder, and Stockton had asked the Secret Service to examine laboratory notes, most of which had been subpoenaed from the authors. We learned of this accidentally—and only a couple of weeks before the next hearing. Again, our source of information was a newspaper reporter, John Crewdson of the *Chicago Tribune,* who warned some scientists that they should not defend us because the Secret Service had discovered incriminating evidence showing that the data had been tampered with.

We were, of course, annoyed that the involvement of the Secret Service was not shared with us, but was known to a small number of newspaper reporters. Further, Crewdson's reaction was typical of the "guilty until proven innocent" attitude displayed by some reporters and by some of those directly involved in the investigation. The mere mentioning of the words "Secret Service" had terrifying import, and carried the implications that we must have done something awful to warrant its involvement. We also started to hear rumors that the O&I subcommittee would soon hold hearings (official notification came just a week before they began).

In the week prior to the hearing, the subcommittee and the Secret Service did allow us to see some of their findings, although they gave us no written report (and, to this writing, no report has appeared). Only a few days before the hearing did we get a few sheets of outlined results of their inquiry. Once we saw the nature of their analysis, it was evident that they had uncovered no secrets and no surprises: They had found that experimental notes were not in sequence, implying poor maintenance of records from experiments. They also found that one illustration was a composite of several photographs. The way the composite was represented by the Secret Service, it sounded as though this accepted and common practice was meant to be misleading, which it wasn't.

A simple inquiry to the authors would have elicited the information it took them nine months of high-tech analysis to gather. Their failure to ask us created an atmosphere in which it seemed that we had something to hide, which we didn't. But a news reporter learning of this development, and not having a period for reflection and further inquiry, could believe that the Secret Service report had uncovered a smoking gun. We had to help educate the press before the hearing, but we did not want to be the source of a leak.

True to form, the subcommittee staff had apparently started a leak; we deduced this because we had earlier heard from others about the material we'd been shown. Thus we felt free to discuss the Secret Service information with the press before the hearing. This was very useful because when reporters heard both sides, they could put things into perspective.

Going into the hearing, we decided that because the charges against us were voluminous and constantly shifting, we needed to individualize our defenses. I concentrated on the effect such congressional reviews might have on science. Dr. Imanishi-Kari, whose work was under direct attack, responded in a personal manner. She focused on the Secret Service findings implying that she was a messy note-taker, which she readily admitted. She also addressed the question of motive, and in dramatic fashion: What possible reason, she asked the congressmen, could she have to cheat? The most frequent citations of the *Cell* paper were by medical researchers investigating autoimmune diseases, particularly lupus. This often-fatal disease took the life of her sister, and Imanishi-Kari herself, she revealed, is likewise afflicted. If she were to deceive lupus researchers, it would be self-destructive.

Who Judges Science?

In my defense, I tried to show how this sort of procedure could seriously harm the functioning of the remarkably successful American biomedical enterprise. One of the reasons that biomedical research has progressed so well in this country is that we have an efficient and effective system based upon peer review. Scientists receive grants only after their proposals are reviewed by other scientists who are expert in their field. The results are likewise reviewed by peers, first when the paper reporting on the work is offered to a refereed journal, and later, after publication, when other scientists verify the results.

This verification process is the cornerstone of American science, which Congressman Dingell, as the son of one of the congressmen who helped to found NIH and brother of a biomedical researcher, clearly wants to protect (while also ensuring that the public gets its money's worth). I would like to help him in this effort, and I am sure that most other scientists would also like to help in an honest examination of the process by which science is reviewed and verified. But fundamentally, I do not believe it is possible to evaluate science in any other way.

The peer-review system came under attack from the moment O'Toole took her accusations outside the university-NIH review process and handed it over to Stewart and Feder, and then on to Stockton. These three unqualified outsiders tried to impose their own review upon the scientific ones that had preceded, even suggesting that the ways in which scientists take their notes need to be regulated to make "auditing" such as their own more convenient. If the sad history of this investigation demonstrates nothing else, it shows that uninformed or malinformed outsiders cannot effectively review the progress of scientific activity.

In a more appropriate manner, our immunological study was followed up with complementary work by, among others, Professors Leonard and Lenore Herzenberg of Stanford University, who found as we had that many cells in the experimental system we studied did not express the transgene. They also found "double-producers," cells that expressed the transgene as well as the "natural" gene, which we did not find. There are reasonable explanations of this discrepancy, a type of refinement that often occurs in science—indeed, it is at the very heart of the scientific process. But it is important to keep in mind that without our study, the Herzenbergs would have been unlikely to do theirs; in turn, their work amplified ours.

The Nature of Collaboration

Another pillar of American science affected by these proceedings was collaboration. Questions were raised repeatedly as to whether I as a coauthor had responsibility for the work performed in a collaborating laboratory and whether I had faithfully discharged that responsibility.

When two laboratories decide to collaborate, two or more professors make the decision and maintain the formal link. They communicate periodically to check on progress, new ideas, or new directions, and from time to time there may be joint meetings of the whole team to review the research. But it is the day-to-day contact between trainees, who are supervised independently, that generally makes the science progress. When writing up results, this network of interactions continues while the draft manuscript goes back and forth, and the final manuscript is an amalgam.

Like many other researchers, I have been involved in numerous scientific collaborations, which are essential for bringing complementary skills to bear on important problems. Trust is at the heart of successful collaboration, and this one was no different. If the scientists in a collaboration are knowledgeable enough to judge the details of each other's experiments, there is little purpose in collaborating. I fear that the type of investigation to which we have been subjected, because of its intimidating nature, could make scientists more wary of collaborating; it could thus undermine an important method for advancing knowledge.

Error Versus Fraud

The errors that have been identified in the *Cell* paper were inconsequential to the conclusions; had this paper not become the focus of such intense scrutiny, the errors would not have been noticed by anyone other than perhaps immunologists seeking to expand upon the results. But now, because of the spotlight, the nature and role of errors have become of increasing concern.

In this connection, it is important to remember that no study is ever complete. The last word is never written. Investigators make the somewhat arbitrary and personal decision to write a paper when they believe that a story—one that makes sense and that others will want to read and build on—can be told. The scientific literature is a conversation among scientists; we describe to each other what we have done, and we implicitly ask our colleagues to try to build on our results, testing their value and their ability to support a growing edifice.

In fact, a scientist's first response upon reading a new paper may be to see a new interpretation, to conceive a different experiment that could provide a wholly new thrust for the paper. This is a thought process that is encouraged in all young scientists, and it is one of the driving forces in science.

I should emphasize that the errors of which I speak must be distinguished from fraud, which involves the conscious misrepresentation of data or the conscious use of data that do not emanate from a described experimental setup. The word "conscious" is crucial here because *un*consciously either of these forms of error could be made and should not be considered fraud. If as a result of this type of investigation American scientists become afraid to make a mistake, then all of science will be slowed, to everyone's detriment.

Policing Science

None of the issues discussed here stands alone, and certainly the policing of science, which means the policing of *scientists*, is related to all the others. Although I believe that errors—and even purposely altered data—will ultimately be discovered by scientific peers, there is obviously the need for an alternative mechanism: One scientist who suspects wrongdoing on the part of another should be able to take appropriate action.

We are currently developing a mechanism at the Whitehead Institute that will go beyond the requirements of a draft regulation issued by the U.S. Department of Health and Human Services. At this point, the procedure is as follows:

- A question about the possibility of scientific misconduct can be raised by anyone in an entirely confidential manner.
- Once a question has been raised, the director appoints a committee of inquiry, composed of appropriate and knowledgeable people; selections are made confidentially and with consideration of real or apparent conflicts of interest. The committee must conclude its work within 60 days, submitting documentation and a written report.
- If evidence of misconduct is found, the director immediately names a committee of investigation to conduct a thorough and authoritative review of the evidence within 120 days. On the strength of the committee's recommendations, the director can take action ranging from reprimand to termination of employment and tenure. Likewise, if the committee finds that the complaint was intentionally dishonest and malicious, the director can take similar action against the accuser.

- Both at the outset of an investigation and after it has been concluded, funding agencies are fully informed.

I believe that the developing Whitehead policy adequately protects the rights of both accused and accuser while providing an effective and speedy procedure for dealing with fraud or misconduct. The policy delineates in no uncertain terms the duties of each person in the Institute for upholding ethical standards, and it keeps adjudicatory responsibility in-house, where it belongs.

Because NIH has ultimate responsibility for the use of its grant funds, it too needs a mechanism for considering charges of misconduct. NIH should determine whether proper procedures have been followed, but not second-guess appropriately constituted reviews. It should also decide what monetary restitution is needed. If it finds that a local review was procedurally faulty or if it believes that only a national review can be truly objective, it should then take over the process.

If the requests of whistleblowers for reviews are kept secret, and the proceedings and verdict published only if significant error or fraud is found, protection of the rights of the accuser and the accused might be more likely. This protection is built into the Whitehead policy, and it should apply universally. Without it, others could be subject to public attacks of the sort directed against me by Stewart, Feder, and Stockton (who need to reflect on the meaning of due process and the American tradition of protecting the rights of the accused). But unlike me, others may lack the resources or the resolve to successfully fight back. They deserve protection. The progress of science depends upon it, and American society should insist upon it.

The scientific community needs to deal promptly and severely with cases of fraud. We have often been loath to do so, perhaps because as honest men and women we not only don't know much about it but are uncomfortable contemplating it. The growing belief that this attitude must change could be a positive result of the publicity surrounding the *Cell* paper investigation.

Scientific Whistleblower Vindicated

Pamela Zurer

Immunologist Margot O'Toole is delighted to be back in the lab. For four years she sat on the outside, branded as the malcontent whistleblower who had dared cross Nobel Laureate David Baltimore.

"I love it," she tells C&EN of her research position at the Genetics Institute of Cambridge, Mass. "I had been left with such a bad taste that I wondered how I'd feel. Now I remember how much I'm stimulated by science."

O'Toole cut a lonely figure in May 1989 as she sat by herself at the long witness table in the hearing room of Rep. John D. Dingell's Energy & Commerce Committee. She had come to explain why three years before she had challenged the validity of a paper by Baltimore, Tufts University immunologist Thereza Imanishi-Kari, and four others (C&EN, May 22, 1989, page 27).

The paper reported that a foreign gene (transgene) inserted into mice influenced the activity of native genes that resemble the transgene [*Cell*, **45,** 247 (1986)]. But, O'Toole testified under oath, "It presented evidence that simply did not exist." The phantom data, crucial to the central claim of the paper, had been contributed by her former boss, Imanishi-Kari.

Five days earlier, Imanishi-Kari, Baltimore, David Weaver (another coauthor), and their respective lawyers had crowded the same table. In their testimony they had rejected O'Toole's concerns as trivial and irrelevant.

That first day the hearing room was jammed with noted scientists come to lend their support to Baltimore, then head of the Whitehead Institute for Biomedical Research and now president of Rockefeller University. Dingell (D.-Mich.) had convened the hearing before his Subcommittee on Oversight & Investigations to examine "the ability and will of major research institutions . . . to police themselves when concerns are raised about potential misconduct."

Baltimore's supporters had rumbled appreciatively as he characterized Dingell's interest as ominous meddling. "I have a very real concern that American science can easily become the victim of this kind of government inquiry," he said. "Professor Imanishi-Kari is also a victim."

O'Toole had no such cheering section as she spoke of her attempts to get the scientific literature corrected. She told how her concerns had been brushed aside by cursory inquiries at Tufts and Massachusetts Institute of Technology, where she and Imanishi-Kari had worked at the time. She related how she had been unable to find a job in science since Imanishi-Kari has kicked her out of the lab. "My competence and motives have been attacked by scientists from all over the world," she said.

Now the tables have turned.

An investigation by the National Institutes of Health's Office of Scientific Integrity confirms O'Toole's allegations. OSI's draft report concludes Imanishi-Kari's actions "constitute serious scientific misconduct." It calls O'Toole a hero.

The supposedly confidential draft report is still under review and may yet be revised or corrected. Sent to the parties involved in mid-March, it promptly was leaked to the press (C&EN, March 25, page 5). As news of its damning conclusions circulated, Baltimore asked *Cell* to retract the paper.

In regard to Baltimore, OSI finds his statements "extraordinary. They are all the more startling when one considers that Dr. Baltimore, by virtue of his seniority and standing, might have been instrumental in effecting a resolution of the concerns about the *Cell* paper early on, possibly before Dr. Imanishi-Kari fabricated some of the data later found to be fraudulent."

Until the report was leaked, Baltimore had steadfastly maintained that nothing of significance was wrong with the paper. In a 1988 letter to scientific colleagues, Baltimore objected to the Dingell subcommittee's involvement as "all totally unnecessary." At the 1989 hearings, he continued to back up Imanishi-Kari, saying he was "a great fan of hers."

Afterward, Baltimore's version of events continued to be circulated widely. He wrote in an article in *Issues in Science & Technology* that "If the sad history of this investigation demonstrates nothing else, it shows that uninformed or malinformed outsiders cannot effectively review the progress of scientific activity." Scientists deluged Dingell with letters supporting Baltimore.

Baltimore's statement sidesteps the fact that Dingell had unveiled much of the incriminating data in the NIH report at the 1989 hearing and at another held in May 1990 (C&EN, May 21, 1990, page 6). Secret Service agents testified that their forensic analyses of Imanishi-Kari's laboratory notebooks showed she had fudged them.

The agents examined the ink on the pages, the paper itself, and gamma-counter tapes that Imanishi-Kari had cut and pasted into her notebooks, binders full of loose-leaf pages. They concluded pages purportedly prepared in 1984 and 1985 were actually created in 1986, about the time O'Toole first raised questions.

As the NIH report puts it, "Imanishi-Kari's notebooks had not been compiled contemporaneously with the conduct of the reported experiments. Rather, the notebooks were assembled specifically to respond to the challenges to the paper."

Baltimore, when asked during the May 1989 hearing for his reaction to the Secret Service's revelation that Imanishi-Kari had been changing dates, responded, "What about it? . . . The way that those notes look to me is not a surprise in the slightest."

The NIH report now reveals that Baltimore and his attorney, Norman Smith Jr., not only knew she had put her notebook together after the fact, they advised her to do so.

A large part of the draft NIH report concerns what OSI calls "the June subcloning data." These are data that convinced an earlier NIH investigation that Imanishi-Kari's work had experimental support, even though the data published in the *Cell* paper did not hold up to scrutiny.

As a result of the first NIH investigation, Baltimore and Imanishi-Kari published the June subcloning data in 1989 as a correction to the *Cell* paper. But it is just those subcloning experiments that O'Toole has consistently said were never done and that the new NIH investigation confirms were fabricated.

Baltimore, however, asserted to OSI that if the June subcloning data were fabricated, NIH was somehow responsible. According to partial transcripts of interviews included in the report, Baltimore said that "in my mind you can make up anything that you want in your notebook, but you can't call it fraud if it wasn't published. Now, you managed to trick us into publishing—sort of tricked Thereza—into publishing a few numbers and now you're going to go back and see if you can produce those as fraud."

O'Toole says Baltimore had ample evidence that the June subcloning data were suspect before the correction was published. "They had my written statement saying Imanishi-Kari had told me those experiments had never been done," she tells C&EN. "And I was sitting with her in June 1989 when she told him they were not done."

O'Toole believes it was not important to Baltimore to correct the *Cell* paper. "He didn't care to correct it," she says. From her point of view the attitude of Baltimore and the other scientists who stood by him unquestioningly is, "Just because the paper was wrong was no reason to correct it. Don't retract stuff—just wait for some poor fool to repeat it and find out it's wrong."

"The thing most upsetting to me," she says, "is the contempt they held for the labor of the people trying to repeat the work. I felt it acutely because I had worked for a year trying to do just that."

Retraction: Altered Repertoire of Endogenous Immunoglobulin Gene Expression in Transgenic Mice Containing a Rearranged Mu Heavy Chain Gene

The undersigned four authors wish to retract the article by Weaver et al. (Cell *45*, 247–259, 1986) because of questions raised about the validity of certain data in the paper. Two authors (Thereza Imanishi-Kari and Moema H. Reis) do not believe that the questions raised have merit and are not parties to this retraction.

David Weaver,* Christopher Albanese, Frank Costantini,† and David Baltimore‡

*Division of Tumor Immunology
Dana-Farber Cancer Institute
Boston, Massachusetts 02115
†Department of Human Genetics and Development
Columbia University
New York, New York 10032
‡The Whitehead Institute
Cambridge, Massachusets 02142
and The Rockefeller University
New York, New York, 10021-6399

Writing Assignments

1. Write a summary of David Baltimore's view of the controversy in "Baltimore's Travels." Describe what he presents as the main issues.

2. Write an essay that considers Baltimore's views in "Baltimore's Travels" in light of what you learn from reading "Scientific Whistleblower Vindicated." What remains valid in Baltimore's article? What is called into question? Why do you think Baltimore eventually retracted the research paper in *Cell*? What are the reasons the authors give in Weaver, et al., "Retraction"? Do you think these are the only reasons?

3. Take a position on the desirability of government investigation of scientific fraud and dishonesty. Write a short essay that seeks to persuade another student to take your position.

3

Reading and Writing to Learn Chemistry

Chemistry has a reputation as a difficult subject. If you entered your chemistry course with the worry that you will not do well in it, you are surely not alone. In this chapter, we will look at some of the reasons students find chemistry difficult and how you can use reading and writing to help you study chemistry.

WHY DOES CHEMISTRY SEEM DIFFICULT TO SO MANY STUDENTS?

Let's take a look at the nature of chemistry and try to determine why it seems difficult to so many students. As you will see, some of the problems students experience learning chemistry are directly related to how chemists work. For this reason, understanding how chemists think about their work can help you cope with the problems chemistry may present.

Problem 1: Finding What's Significant

The first problem is that chemistry is based on thousands and thousands of observations. For example, a chemical observation from the ancient world was that the solid yellow substance called sulfur burned in air with a blue flame to yield a gaseous product that made people almost cough their lungs out. A recent example is the discovery by Richard Smalley and coworkers at Rice University in Texas in 1985 that there exists a new form of carbon with atoms arranged like the seams of a soccer ball. Over the last two hundred years the progress made in chemistry has been astonishing.

How many of these observations do you need to know? Experienced chemists know which observations are significant for their work, that is which ones they need to know well to carry out their research. This isn't the case, though, with students who are just learning the field. Instead, you may feel overwhelmed by a bewildering number of observations. Therefore, one problem in learning chemistry is sorting through the considerable number of known facts, evaluating their significance, and learning which ones are important.

Problem 2: Understanding Models

Learning chemistry, however, is not just a matter of memorizing observations. The second step in chemistry, after the facts have been collected, is to try to assemble a suitable *model* that explains what has been observed. This is probably the part that chemists enjoy the most, and it is probably what makes chemistry difficult for many students.

The problem is that in chemistry the actual occurrences take place at the level of atoms and molecules, items so small that they cannot truly be seen. We call this the *microscopic* level. The facts or data that we collect are always at the regular scale of observable events—substances burning or boiling or melting, for example. We call this the *macroscopic* level.

Thus chemists want to link the macroscopic observations we know about with a microscopic model. A model that is consistent with all of the data in an area of chemistry is considered successful and becomes accepted as a working hypothesis for further study. Dalton's theory of atoms and molecules or Watson and Crick's double helix model of DNA are early and recent examples.

The problem of understanding chemical models is that of envisioning what goes on at the microscopic level and how it relates to the macroscopic level. The distances between atoms and the masses or weights of atoms are incomprehensibly small and the numbers of atoms even in small samples of matter are incomprehensibly large. The number of atoms in a speck of sand that you can barely see is many, many times greater than the number of all the people, past and present, on the Earth. The ordinary rules of behavior that work for macroscopic objects such as baseballs or grains of sand have to be supplemented by new rules for the microscopic atoms, molecules, and electrons.

Furthermore, our model often has to shift when we are looking at different properties of the same object. The most famous example of this is that small particles such as atoms and electrons can behave as moving particles or as waves similar to light waves. Well, which one is right? Does an electron behave like a particle or a wave? The answer is that it behaves like neither. An electron behaves like an electron and the words particle and wave are just terms we are familiar with that approximate the behavior of an electron under certain circumstances. Thus we must expect that understanding these different models can be complicated.

Problem 3: Reasoning by Analogy

It would be wrong to suggest that all chemists do is construct models to explain nature. Chemistry is a highly practical subject that is useful in predicting how matter will behave. This is generally done by *analogy,* a form of reasoning in which known similarities are used to infer unknown properties. The typical problem facing a chemist in industry or a student in an exam is to figure out how to produce a new product or explain an unfamiliar phenomenon by drawing on the chemical facts already known.

A chemist in industry asked to develop a new high temperature plastic, for example, would do so by drawing on a store of knowledge on existing plastics and the compounds and conditions used to produce them. For a beginning chemistry student the task might be to predict the properties of H_2S. In this case, the student may be quite familiar with the compound H_2O and can then construct a reasonable picture of the nature of H_2S by using what he or she knows about the similarities and differences between oxygen and sulfur.

In these instances, reasoning by analogy is an essential part of chemistry, whether the task is to develop a new product or to answer a question on an exam. But in both cases, reasoning by analogy involves moving from the known to explain the unknown, and this may be a form of thinking with which some students are unfamiliar.

Exercises

1. List the models that are used to explain chemical phenomena in one chapter of your chemistry textbook.
2. What do you see as potential problems in producing and using models to explain chemical phenomena? Give examples to support your view.

3. Give a few examples of reasoning by analogy in chemistry or another subject.

READING YOUR TEXTBOOK

As soon as you start your first few reading assignments in chemistry (or any other science) you quickly come to the realization that it is different than reading in nonscience subjects. The reading assignments in chemistry are not nearly so long; you never hear a student say that she has a 200-page reading assignment in chemistry this week. However, the chemistry assignment may be just as hard as that 200-page assignment in, say, history.

The nature of science writing makes it slow to read because so much information can be packed into a very few words. This means that when you read just a few paragraphs you may encounter a number of important concepts, all of which take thought and effort to be fully understood. Thus the first principle in reading an assignment in chemistry is to *slow down* so that you can concentrate on the density of information being presented.

Exercise

The following selection gives an example of how compact the material is in a chemistry text. This excerpt is a short section from the second chapter of a popular general chemistry text designed for freshman college students who are nonchemistry majors intending to enter health-related fields such as nursing, nutrition, and medical technology (Bettelheim and March, *Introduction to Organic & Biological Chemistry*). As an exercise, read this excerpt through quickly and then write down from memory what you recall of the concepts that were presented.

> ***Chemical and Physical Changes***
>
> It has long been known that matter can change, or be made to change, from one form to another. In a **chemical change**, substances are used up (disappear) and others are formed to take their places. An example is the burning of propane ("bottled gas"). When this chemical change takes place, propane and oxygen from the air are converted to carbon dioxide and water. Chemical changes are more often called *chemical reactions*. Many thousands of them are known, and the study of these reactions is the chief business of most chemists.

Matter also undergoes other kinds of changes, called **physical changes**. These changes differ from chemical reactions in that substances do not change their identity. Most physical changes are changes of state—for example, the melting of solids and the boiling of liquids. Water remains water whether in the liquid state or in the form of ice or steam. Conversion from one state to another is a physical, not a chemical change. Another important type of physical change involves making or separating mixtures. Dissolving sugar in water is a physical change.

When we talk of the *chemical properties* of a substance, we mean the chemical reactions it undergoes. *Physical properties* are all properties that do not involve chemical reactions; for example, density, color, melting point, and physical state (liquid, solid, gas) are all physical properties.

After you have jotted down the concepts you recall from your first quick reading of this excerpt, go back through it again, reading carefully and slowly, and underline every important concept. Then go back through again and list every concept that is presented. Check your list with the list below. The number of concepts presented in this short excerpt may surprise you.

The list of concepts we compiled from this excerpt is:

1. A chemical change is a process in which some substances are used up and other, different substances are formed.
2. Combustion is an example of a chemical change.
3. A chemical reaction is just another name for a chemical change.
4. Studying chemical change is the essence of chemistry.
5. Physical changes do not involve the change of identity of any substance.
6. Melting and boiling are called changes of state.
7. Changes of state are physical changes.
8. Forming or separating mixtures involves physical change.
9. Dissolving one substance in another is an example of the formation of a mixture and is a physical change.
10. The chemical changes or reactions that a substance can undergo define its chemical properties.
11. Physical properties do not involve chemical reactions and therefore do not involve changing the identity of the substance.

Clearly this passage is very compact and involves a lot of concepts. Therefore, it cannot be read for content rapidly. The trick to reading chemistry is keeping your mind on the reading and not allowing a crucial concept to slip by. It is usually wise to realize that you can only digest a limited number of concepts at one time. Therefore, don't sit down to read too much at once. The following reading and writing strategies offer some ways to deal with the compactness of science writing and to help you understand and learn the material presented.

Get an Overview of What You're Reading

If you are assigned a chapter for homework, begin by skimming through the whole chapter, noticing how many separate sections it has. Read the introduction to the chapter to get a general sense of what the chapter is about. Read the section headings to see what topics the chapter treats. Set realistic goals for yourself by dividing the chapter into its sections and then reading them carefully, one at a time.

Assess Your Understanding

During your reading it is important to keep assessing whether you are really understanding and incorporating the concepts as you are reading. One way to do this is to read your assignment a section at a time. Wherever you see a section heading, such as "Chemical and Physical Changes" in the preceding selection, convert the section heading into a question ("What are chemical and physical changes?") and read that section in order to answer the question. Once you have finished the section, pause for a moment and see if you can answer the question without looking at the textbook. Check your answer by reviewing the section.

The way that one chemical concept builds on another, particularly as presented in textbooks, means that you need to understand each concept before going on to the next section.

Read Actively by Underlining, Writing in the Margins, and Listing Key Concepts

Reading with a pen or pencil in your hand is a good way to make your reading more active. *Underlining* or highlighting is useful, but it must be done in a meaningful way. If you have effectively underlined a section in your book, when you go back through that section you should be able to go from one underlined sentence to the next and clearly see that you are capturing the main concepts. If you have underlined everything, you are probably not picking out key concepts and points as well as you might. We have underlined the passage "Chemical and Physical Changes" to give you an idea of how you might catch the key points.

> ***Chemical and Physical Changes***
>
> It has long been known that matter can change, or be made to change, from one form to another. In a chemical change, substances are used up (disappear) and others are formed to take their places. An example is the burning of propane ("bottled gas"). When this chemical change takes place, propane and oxygen from the air are converted to carbon dioxide and water. Chemical changes are more often called chemical reactions. Many thousands of them are known, and the study of these reactions is the chief business of most chemists.
>
> Matter also undergoes other kinds of changes, called physical changes. These changes differ from chemical reactions in that substances do not change their identity. Most physical changes are changes of state—for example, the melting of solids and the boiling of liquids. Water remains water whether in the liquid state or in the form of ice or steam. Conversion from one state to another is a physical, not a chemical change. Another important type of physical change involves making or separating mixtures. Dissolving sugar in water is a physical change.
>
> When we talk of the chemical properties of a substance, we mean the chemical reactions it undergoes. Physical properties are all properties that do not involve chemical reactions; for example, density, color, melting point, and physical state (liquid, solid, gas) are all physical properties.

Writing in the margins of your textbook is another way to help you think about the material as you read—write notes to yourself about important concepts, how the ideas relate to each other, and what points are confusing.

Listing the important concepts as you go is a good exercise to keep your mind on the concepts while reading and also serves as a useful guide when reviewing in the future.

Write Chapter Summaries in Your Own Words

One way to make sure you understand what you have read is to summarize the material covered by a chapter in your own words.

Concepts that seem clear on the printed page may slip away when you put the book down because they are described in someone else's words. Learning occurs when you can put these concepts in your own words. You may be able to memorize a definition or a formula, but it's only when you can explain the definition or formula to yourself, in your own words, that you have made it your own.

When to Read: Before or After Lectures?

There are obviously different strategies for reading assigned material. Usually, the professor will give the assignment with the intention that you will read the material before she or he covers it in a lecture so that you will understand the information better when it is presented in class. Whether you actually read the material before class is clearly up to you. There is no absolute rule as to whether it is better to use the reading beforehand to help you understand the lecture or to use the lecture to help you understand the reading that you do after the lecture. You could even combine techniques and read the text both before and after the lecture.

However, there are two realities that must be faced. First, as a student your time is limited. Second, a quick skim through the reading will be useful only as a preliminary to the real study of the material. Therefore, you have to find out what works best. You should try different patterns to see which is best for you. If reading before the lecture works best, do it that way. If something else works better, then change. A cardinal rule of studying chemistry (or any other subject) is to ask yourself frequently, "*Is this mode of studying working for me?*" Putting in a lot of time studying does not necessarily guarantee that you will get a good grade. You have to choose studying patterns that work for you, and only you can tell which are your successful patterns.

Doing Problem Sets

In chemistry, reading assignments are normally given with pertinent problems from the end of the chapter to test your understanding of the concepts covered. Some students do not read the assigned material but go right to the problems and try to solve them. When they cannot solve a particular problem, they go back into the chapter and find a worked example that is similar enough to the assigned problem to provide the necessary parallel steps. This method of using the textbook neither saves time nor gives the best preparation for exams. It simply treats the problem sets as homework to turn in and won't help you understand the concepts you need to know.

Pay Attention to Formulas, Equations, Graphs, Tables, and Illustrations

Reading in science in general and chemistry in particular is often interrupted by references to diagrams, graphs, and tables and by the use of formulas and equations, both chemical and mathematical. These break the flow of the reading and each of these interruptions provides a point where you can lose the thread of the reading. For many of us, diagrams and equations are a less familiar form of written communication than regular text. Thus we often skim across these features without completely comprehending them. However, in chemistry diagrams and equations are usually essential to the development of ideas. For this reason, it is important that you learn to read them as closely as you read regular text.

We mentioned above that underlining and writing are useful tools for reading a textbook. However, underlining or highlighting each equation as it appears becomes mechanical and is unlikely to aid in comprehension. On the other hand, writing out the equations to explain what they mean and performing the mathematical manipulations described can be quite useful to you. Drawing the essential parts of a diagram can also be useful. Remember that when you see a diagram, a formula, or an equation that you should expect to spend more time with it than regular text, not less time. That is because diagrams, formulas, and equations are usually shorthand methods of expressing ideas.

Notice in the following passage from a general chemistry textbook, Fine and Beall, *Chemistry for Engineers and Scientists,* how the written text and formulas work together to explain and illustrate Avogadro's hypothesis in shorthand form.

> ***Avogadro's Hypothesis: The Mass and Volume of a Gas***
>
> The essential features of a trapped gas (Figure 5-5) can be seen in the relationships between temperature, volume and pressure. **Avogadro's hypothesis** states that equal volumes of gases at the same temperature and pressure contain the same number of molecules. Because this simple statement holds, regardless of the identity of the gas, we are led to important generalizations about gases.

If an equal number of molecules of all gases have the same volume at the same temperature and pressure, then (as long as temperature and pressure are held constant) the volume of a gas will be proportional to the number of molecules it contains. Since the number of molecules contained in even the smallest measurable amount of gas is enormous, we choose a convenient standard, the mole—a mass of gas equal in grams to the formula weight, or 6.022×10^{23} molecules. Thus Avogadro's hypothesis can be stated,

$$V_{T;P} = k \cdot n \quad (5\text{–}2)$$

V is the volume of the gas and n is the number of moles. The subscripts T and P specify that the temperature and pressure are held constant. The quantity k is a proportionality constant; its value depends on the temperature and pressure and has units of volume/moles. Now, the number of moles n of a gas can be calculated from,

$$n = \frac{m}{M}$$

where m is the mass of the gas and M is its molecular weight. Substituting into Eq. 5–2 gives,

$$V_{T;P} = \frac{k \cdot m}{M}$$

which can be rearranged to,

$$\frac{1}{k} M = \frac{m}{V_{T;P}} = d$$

where mass divided by volume is equal to the density d of the gas. Now, since $\frac{1}{k}$ is a constant, we can conclude that **the density of a gas is proportional to its molecular weight,** another important consequence of Avogadro's hypothesis.

Exercises

1. Underline or highlight the most important parts of one section of a chapter in your textbook and have another student do the same. Then compare the two sets of underlining and comment on the similarities and differences.

2. Write a list of the key concepts in a chapter in your textbook and have a classmate do the same. Compare the lists. How well do they match? Discuss the similarities and differences.

3. Write a summary of a chapter in your textbook and have a classmate do the same. Compare the summaries. How are they alike and different? Don't try to judge whether one is better than the other. The point here is to consider the strengths and weaknesses of each.
4. Take a paragraph in your chemistry textbook that contains chemical or mathematical equations and rewrite the paragraph in words only. How does the length of the paragraph change?

TAKING AND USING LECTURE NOTES

In chemistry, as in many subjects, lectures are the central part of the course. This is the place where the teacher emphasizes the concepts and gives the explanations that he or she considers the most important. These points are likely to return in quizzes and examinations. Almost everyone, students and teachers alike, is convinced that taking lecture notes is an essential part of being successful in a chemistry course.

In fact, taking lecture notes is one of the most important forms of writing you will be called on to do in a chemistry class. Often, students and teachers don't think of taking lecture notes as a kind of writing assignment. It's just what students do during a lecture. But this view of lecture notes misses an important point. Taking good notes and using them effectively is a form of writing that students need to—and can—master. In this section, we offer some suggestions about how you can more effectively take notes in order to learn chemistry.

Use Note-taking as a Way to Follow the Lecture

The student's task in taking lecture notes is clear if not easy—write like crazy, sorting out the major points as you go, so that you end up with notes that will be intelligible and useful when you return to them later. The real question here is how you can do all this at one time. If you try to write everything down, it may be difficult to distinguish between major and minor points as you're writing.

The first advice we offer is to think of yourself taking notes in a lecture not as a recording instrument but as an intelligent listener.

Use note-taking as a method of keeping your mind on the main thread of the lecture as it is being given. Develop your own system to highlight major points, such as underlining, circling, and starting a new line.

Note Where you get Lost or Feel Confused

Your notes should enable you to reconstruct not only the major points of the lecture but also how the points are connected to each other. When you are taking lecture notes, observe when your note-taking gets "on the wrong track"—when you find yourself having written a couple of pages that do not seem to follow the point that the lecturer is making—or when you feel confused. Make a note in the margin here so that you can return to this section later to clarify the concepts presented.

Some lecturers have a talent for organizing their lectures in a way that seems intuitively logical to students while the presentation is in progress. It will be easier to take notes during such lectures. In other cases, the lectures may not be as well organized, and they will be harder to follow. If this is the case in your chemistry lecture, you'll need to make notes to yourself about sections of the lecture that seem confusing or unclear.

Notice the Relationship Between What the Lecturer Says and Visual Aids and Demonstrations

In addition to lecturing, many chemistry professors use teaching aids, such as the blackboard, overhead transparencies, videos, computers, and, of course, lecture demonstrations. Some students write down everything that the lecturer puts on the board, transparency, or video monitor and just listen to everything else. Copying such material, however, will not capture the lecture content. Your task is to relate what the lecturer presents on the blackboard, transparency, or video monitor to the main points in the lecture.

Review Your Notes After the Lecture

It is clearly important to concentrate on getting the main points of the lecture captured in your notes. Probably the best way of doing this is to review your notes as soon as possible after class while the words of the lecturer are fresh in your mind. At this point you can evaluate how well you did in the note-taking and add any items that come to mind.

The strength of your lecture notes is that *you* took them and as such they have special meaning to *you*. Reading your own lecture notes with your own words and your own conventions and abbreviations is quite different than plunging into your chemistry text, every word of which was written by somebody else. You need to realize, however, that the benefit of personal familiarity will decrease and the problem of questionable organization will increase with the time between when the notes were taken and when they start getting used. Even though it may be a natural thing to want to put the lecture and the notes out of your mind for a while after the lecture, it is best to get back to the notes as soon as possible.

We have already suggested that you review the notes while you still can recall the lecture and you are able to make additions and corrections. Even better at this stage is to rewrite your notes completely now that you have the organization of the total lecture in your mind. Another useful technique is to write a summary of its main points. Exactly what you do with your lecture notes is up to you. The most important point is that you do it right away.

Exercises

1. Rewrite the notes of a chemistry lecture the evening after the lecture. What did you learn from this experience?
2. Compare your notes for a chemistry lecture with those taken by another student. Comment on the similarities and differences.

STUDYING FOR QUIZZES AND EXAMINATIONS

The most important rule of studying chemistry is *study as you go!* Because of all the interlocking concepts and modes of thinking involved, it is usually impossible to master a lot of material in a short time. If you put off studying until the night before the exam, you will almost surely be unsuccessful. Chemistry is best studied in small doses, giving you time to assimilate difficult ideas.

But you also need to think about what you are studying and what your chemistry professor is likely to test you on. In the first section of this chapter we talked about the nature of chemistry and why many students find it difficult. Of course, these difficulties will really show up in quizzes and examinations where the teacher is going to ask you to "think like a chemist." The tasks you may face in a quiz or examination are:

1. Knowing a number of *facts* and *chemical conventions*. The conventions will include knowing the names and formulas of a number of important compounds and writing the chemical equations for reactions correctly. The facts will include the properties and reactions of important compounds, the nature of important processes, and equations that can be used to make calculations.
2. Understanding the models used to explain the microscopic properties of matter. Your teacher may not even use the word *models,* but as soon as you hear a statement like, "we will consider the atoms to be hard spheres in contact with each other," or "the volume of an *ideal* gas is proportional to the absolute temperature," or "the electrons have a high probability of being found in a certain region around the nucleus," you have a *model.* You should not only know the models but you should have some feeling for their limitations.
3. Being able to reason by *analogy,* that is, being able to use some facts that you know to reason out one or more that you don't know.

The task of studying for exams and quizzes involves sorting out the most important facts and concepts and then incorporating the thinking skills necessary for the discipline. Here are some suggestions about how to do these things.

Memorizing

All students recognize the need for identifying the important facts and committing them to memory. In the sections above we have considered the problem of identifying the most important points. Memorization skills, when you need them, are the same as you would use for any other subject. For example, flash cards or, if available, an interactive computer program are useful for memorizing the names and formulas of inorganic compounds. If there is material you need to memorize, divide it up so that you can work on small sections of material at a time.

"Thinking Like a Chemist"

Too often students overemphasize memorization in studying chemistry and don't devote enough time to learning how to think like a chemist. The more important aspects of "thinking like a chemist" involve understanding models and thinking by chemical analogy.

What are the ways to develop these skills? They are developed by using them, by talking about chemistry. It is wise to use every opportunity to talk to your chemistry teacher, the teaching assistant, other more advanced students in chemistry, and your own peers about the models of chemistry and how they are used.

Study and discussion groups are very effective. Talk about how problems are solved in chemistry, particularly the kinds of problems that you are covering in your course. If possible, have a study group with a more advanced student as a leader. The study group will probably be as useful for the leader as for the less advanced students. Also, you should take the opportunity to explain chemical concepts to another student who does not understand them as well as you do. You will find that in explaining the concepts you will come to understand them better yourself.

Preparing for Different Kinds of Quizzes and Examinations

To be successful in your chemistry class it is also important to consider the nature of the quizzes and examinations that your teacher gives. In chemistry these can be solving numerical problems, short answer questions, multiple choice questions, and, less commonly, essay questions. We will consider tips for preparing for and taking each kind of test.

Numerical problems

For tests containing numerical problems it is, of course, important to have practiced working problems beforehand. *Do not confuse reading through somebody else's solution to a problem with working it yourself.* Somebody else's solution always makes sense when you read it. The trick is being able to solve the problem yourself. You should practice working problems in an orderly way, putting down each step so that you can clearly follow all of the steps you took. Every step should make sense and you should be able to explain why you did what you did. An orderly, methodical solution of the problem on the test will give you the best chance of being correct and will net you the most partial credit if you are not.

Short answer and multiple choice questions

Studying for short answer and multiple choice questions involves identifying the key topics in your course. Study groups can be very useful. You might, for example, make your own list of what you think is important for each section of the course and compare it with the lists other students have made. It is also useful to make a list of the concepts you do not understand and why. Such an exercise can often lead to identification and comprehension of the important ideas.

In taking short answer and multiple choice tests, it is very important to read the questions carefully. Make sure you understand what the question is asking. In short answer questions, pay particular attention to terms such as "define," "compare," "contrast," and "explain." Multiple choice questions test the depth of your understanding by offering answers that are tempting but incorrect.

Essay questions

Essay questions are not as frequent in chemistry quizzes and examinations as numerical problems, short answers, and multiple choice items. To handle an essay question well, you need to manage your time. One useful strategy is to allot one-third of the time available for writing the essay to brainstorming ideas and planning your essay. Read the essay question carefully, making sure you understand what it calls on you to do. Then brainstorm ideas you can use in your essay. Write down a list of key concepts that you think should appear to answer the essay question. Then plan your essay before you start writing. You can sketch a brief outline of your points in the order you want to present them. This will help you to see the connections between the ideas. Spend the remaining two-thirds of the time composing the essay. If you can, leave a few minutes at the end to read over your essay to insert any key ideas you have left out and to correct misspellings and grammatical errors. Make the additions and corrections as neatly as you can. You don't have to copy the essay over.

Check Your Work

Once you have finished a quiz or an exam, use any time left to check and recheck the anwers you have given. If time permits, rework the numerical problems without looking at your previous solution.

Exercises

1. Analyze the following multiple choice questions and answers. What is it about the way the questions are written that might make them tricky or misleading?
   A. Which of the following describes sugar dissolved in water?
      i. An element
      ii. A compound
      iii. A mixture
      iv. None of the above
   B. What is the maximum mass of water that can be prepared from 20.0 g of O_2 and an excess of hydrogen?
      i. 11.2 g
      ii. 22.5 g
      iii. 45.0 g
      iv. This question cannot be answered without knowing how large the excess of hydrogen was.
   C. Which of the following is true of the law of conservation of mass?
      i. It is always absolutely true under all circumstances.
      ii. It is not totally true and therefore it is now useless.
      iii. It is useful for describing almost all chemical processes.
      iv. It is a law of physics and therefore is not used much in chemistry.
   D. What is the formula for calcium nitrite?
      i. Ca_2NO_2
      ii. $CaNO_3$
      iii. $Ca(NO_2)_2$
      iv. $CeNO_3$
      v. $Ca(NO_3)_2$

E. Which of the following statements about chemical compounds is not true?
   i. Two elements cannot combine so as to form more than one compound.
   ii. The composition of a particular compound is fixed.
   iii. A compound is not considered a pure substance since it contains more than one element.
   iv. The smallest unit of a compound is a molecule.

2. Analyze the following short answer questions. In each instance, what is the question asking for? What would a successful answer be?
   A. Contrast *mass* and *weight.*
   B. Define the word *molecule.*
   C. Explain the relationships among the numbers of protons, neutrons and electrons in an atom.
   D. Compare *physical processes* with *chemical processes.*
   E. Describe how the percent yield is calculated for a chemical reaction.

3. Write a test for the section that you are studying now in chemistry. Have a fellow student do the same and then take each other's tests. Comment on how and if this helped prepare you for the actual test.

4

Writing Laboratory Reports

The laboratory is the place where students actually get to do chemistry. The reading and writing you have done to learn chemistry from textbooks and lectures take on a new form in the laboratory—that of the practice of experimental science. In the chemistry lab you learn what it means to investigate chemical phenomena as part of the disciplined inquiry that organizes the community of chemists and gives them their sense of purpose. You learn how to ask questions you can answer by experimental means; how to record and organize results in written, numerical, and graphic forms; and how to interpret results and communicate your interpretations to others persuasively. As you will see in this chapter, laboratory work requires lots of reading and writing—before, during, and after experiments. In this chapter, you will be introduced to some of the writing typically associated with laboratory work—preparing for experiments, keeping a laboratory notebook, and writing laboratory reports.

THE LABORATORY IN THE STUDY OF CHEMISTRY

The following excerpt from a biography of Ira Remsen is an interesting and highly personal example of the learning that can occur in the laboratory. Remsen (1846 –1927) was a pioneering American chemist and educator and established one of the first important instructional laboratories in the country at Johns Hopkins University. He was also president of Johns Hopkins for more than ten years and was a dedicated public servant. In the following excerpt he was still a medical student and not yet in chemistry full time. As you read, notice how Remsen repeatedly contrasts learning chemistry "by reading out of a book" and learning chemistry through experiment and observation.

From Life of Ira Remsen

Frederick H. Getman

Remsen's reminiscences of this period of his student days are both interesting and amusing. His inimitable gift as a raconteur is admirably illustrated by his account in later years of his first attempt to prepare hydrogen. "In those times," said Remsen, "the medical student was assigned to a preceptor, and my preceptor was the professor of chemistry. Under his guidance I had learned much chemistry by reading out of a book. I even assisted in preparing his lectures, and in this work I learned vastly more than from the book, since I came in such close contact with chemical substances, or rather they came in such close contact with me. I well remember when, as the first experiment, he asked me to prepare a bell-jar full of hydrogen. The apparatus provided was an elaborate affair and I had not the slightest idea what to do with it. I saw some activity going on within the apparatus and then I remembered that hydrogen is a gas that burns. I tried to see if what I was producing would burn. A tremendous explosion followed. Everything except myself disappeared, and I could not imagine why I did not go along with the rest." Elsewhere Remsen told the following incident relative to this period of his career: "While reading a textbook of chemistry," said he, "I came upon the statement, 'nitric acid acts upon copper.' I was getting tired of reading such absurd stuff and I determined to see what this meant. Copper was more or less familiar to me, for copper cents were then in use. I had seen a bottle marked 'nitric acid' on a table in the doctor's office where I was then 'doing time!' I did not know its peculiarities, but I was getting on and likely to learn. The spirit of adventure was upon me. Having nitric acid and copper, I had only to learn what the words 'act upon' meant. Then the statement, 'nitric acid acts upon copper,' would be something more than mere words. All was still. In the interest of knowledge I was even willing to sacrifice one of the few copper cents then in my possession. I put one of them on the table; opened the bottle marked 'nitric acid;' poured some of the liquid on the copper, and prepared to make an observation. But what was this wonderful thing which I beheld? The cent was already changed, and it was no small change either. A greenish blue liquid foamed and fumed over the cent and over the table. The air in the neighborhood of the performance became colored dark red. A great colored cloud arose. This was disagreeable and suffocating—how should I stop this? I tried to get rid of the objectionable mess by picking it up and throwing it out of the window which I had meanwhile opened. I learned another fact—nitric acid not only acts upon copper but it also acts upon fingers. The pain led to another unpremeditated experiment. I drew my fingers across my trousers and another fact was discovered. Nitric acid acts upon trousers. Taking everything into consideration, that was the most impressive experiment, and relatively, probably the most costly experiment I have ever performed. I tell of it even now with interest. It was a revelation to me. It resulted in a desire on my part to learn more about that remarkable kind of action. Plainly the only way to learn about it was to see its results, to experiment, to work in a laboratory."

Writing Assignments

1. Describe an instance where you learned something by experience and observation that you might not have learned as well by reading a book. Use this example to explain the differences and similarities of "book learning" and "learning by experience and observation."
2. Remsen seems to be presenting the instructional laboratory as a place necessary for the learning of chemistry. What does Remsen suggest students learn in a laboratory that they can't learn elsewhere? Do you agree with him and why do you feel this way?

PREPARING FOR THE LABORATORY

The laboratory requires you to conduct experiments, make observations, record results, and draw conclusions. It also gives you the opportunity to see in reality what your teacher and your textbook have been referring to. What you get out of the laboratory will depend on how well it has been planned and structured, but it will also depend on how much you put into it. The laboratory is a form of active learning and the extent of your own activeness will determine the pleasure and value that you get from it.

Many students do not prepare adequately for their chemistry laboratories, and so their work in the laboratory is often more time consuming and less rewarding and instructive than it should be. Many laboratory experiments include a pre-laboratory section to make sure that preparations are made. In order to maximize the benefits of your laboratory experience and minimize the time necessary to complete the work, we recommend that you prepare for the laboratory whether or not a pre-laboratory exercise is required. This preparation should include:

1. Reading through the entire experiment, gauging how long procedures should take.
2. Writing down what you think the purpose of the experiment is; this will help in writing the report later.
3. Writing down key aspects of the experiment including concepts, chemical equations, and mathematical equations that you will be using.
4. Making a note of those aspects of the experiment that are not clear to you; this will allow you to clear up your difficulties with the laboratory teacher right away and avoid wasting your laboratory time.

THE LABORATORY NOTEBOOK

As you can see from the Imanishi-Kari case discussed earlier (see Chapter 2, "A Case of Possible Scientific Dishonesty"), in the actual practice of chemistry in the industrial or university laboratory, the laboratory notebook is as important as the most sophisticated instrument. It is here that the practicing chemist records ideas, the reasons for performing experiments, details of procedures, data acquired, tabulations of results, interpretation of observations, and suggestions for further work. Clearly the laboratory notebook contains all kinds of useful information, and going back and reading it can provide even more information than the chemist realized at the time.

The Notebook in the Industrial and Academic Laboratory

Every day in industry, working laboratory notebooks are signed and dated by the investigator and then read, signed, and dated by another scientist who understands the content. Industrial laboratory notebooks are the property of the company and are kept carefully filed.

The notebooks themselves are sturdily bound using expensive acid-free paper. The writing is done with special nonfading ballpoint pens, and archival quality tape or acid-free white glue are used to attach any items such as strip-charts or photographs to the notebook pages. The reason for all of this care is that the laboratory notebook is the principal record that a company has to establish its precedence and right to patent any new discoveries that are made in its laboratories. These patents, of course, are the lifeblood and future of the company. For the individual scientist, the laboratory notebook is the proof of deserving credit for a discovery and is thus essential to advancing his or her career.

The notebook is also essential to performing effective experiments and drawing the most significant possible conclusions in the academic laboratory. An interesting example of the consequences of the failure to keep a good notebook is the discovery of the planet Uranus. The French astronomer LeMonnier recorded observations in 1763 and 1769 that should have told him that he had discovered a new planet. Had he recorded the daily positions of the celestial object that he was observing in regular columns, he would immediately have recognized that it was a planet since it was moving with respect to the stars. Instead, however, he wrote down his observations on random slips of paper, including a paper bag that had formerly held perfumed wig powder. The discovery of Uranus had to wait until March, 1781, for Sir William Herschel who did know how to keep good records and who received credit for the discovery instead of LeMonnier.

Written Laboratory Records

A number of different types of written records can be kept in the chemical laboratory. These include *logbooks*, which are chronological lists of such things as samples processed, instruments used by different people, and equipment breakdowns and maintenance. Each logbook is normally restricted to a narrow activity, and the logbook for a particular instrument is often kept right next to it. Logbooks may be made up of form sheets that are written separately and then collected in chronological order in a binder. The scientist may also keep a *diary* or *journal*, which can include personal opinions and comments on any subject. Keeping such a journal is valuable, particularly as part of a learning process.

The *laboratory notebook* that we are concentrating on in this section is normally written on numbered pages by only one person and is restricted to scientific ideas, facts, and interpretation. In recent years there have been a number of cases of alleged research fraud in university laboratories. The laboratory notebook is the principal evidence in these cases that establishes or disproves charges of scientific wrongdoing. Therefore, the academic laboratory should follow the same standards for laboratory notebooks that exist in industry where their purpose is to justify patents. Academic laboratory notebooks should be permanently bound, written only in indelible ink, have no blank spaces, and be signed and witnessed daily.

Keeping a Laboratory Notebook

Writing up a good laboratory notebook is a skill that is learned by careful practice. Because it contains many different kinds of information including ideas, plans, experimental details, data, results and speculations, there is no set format that can be prescribed. Instead you must be flexible and adjust the format that you use to produce a notebook that will be most useful to you and anyone else who is likely to read it. Here are a few general considerations to take into account about keeping a laboratory notebook:

- All of the pages of the notebook should be consecutively numbered; number the pages in the upper outside corners yourself if the notebook doesn't come that way.

- Create a table of contents at the beginning of the notebook. Each time you make a new entry in your notebook, enter the date and a brief description of the entry in your table of contents. This will give you a running account of what is in a particular notebook and where to find it.

- You can also write a preface stating the purpose of the notebook in a few paragraphs once it is filled. Notebooks build up in number over the years and later it can be useful to have a brief and succinct statement at the beginning of the notebook to help you decide if this is the one you have to go through page by page.

- Create a table of the abbreviations you're using in the notebook. It is undesirable to write out the same long names over and over, but you need a key to abbreviations. Then when you look back at your notebook in a few years you'll be able to tell easily that AIE was your abbreviation for amyl isopropyl ether and that AIL meant analytical instrument laboratory.

The description in your notebook of a particular project should be broken down into the same logical sections that you use to approach the actual work. You might use the following sections to keep a record of a particular experiment or set of experiments:

- Purpose,
- Plan,
- Procedures,
- Results,
- Discussion, and
- Conclusions.

We will now look at the writing of each of these sections showing samples that will come from the same experiment.

Purpose

The Purpose is the first section you write in your notebook on a project that you are just starting. The Purpose section should present a clear statement of the scientific problem you are investigating and why you are doing so. The initial sentence should define the problem. Then follow with an explanation of the proposed work, the reason that it is being done, and the results of previous work that has been done in this area of study, including references, if appropriate, from the scientific literature. A statement of purpose for a laboratory experiment, the variation of gas volume with temperature (Charles's law), is given in Figure 4.1. In this classic experiment it was not appropriate to include references. Note that the unused space at the bottom of the page has been crossed out.

Plan

The Plan is your best estimate of exactly what you will do in this project. This should be as detailed as you can make it at the outset. However, it is unnecessary to describe the work of others fully if it can simply be referenced. Include drawings, tables, diagrams, or other illustrations whenever these are more effective than words alone. Also include the calculations you make in the planning stage of the project, such as how much of each starting material will be necessary. As the project progresses, you will find that you may well do things differently than you expected to do them in the Plan. This section records your original intentions and as such can be very useful. The sample plan is shown in

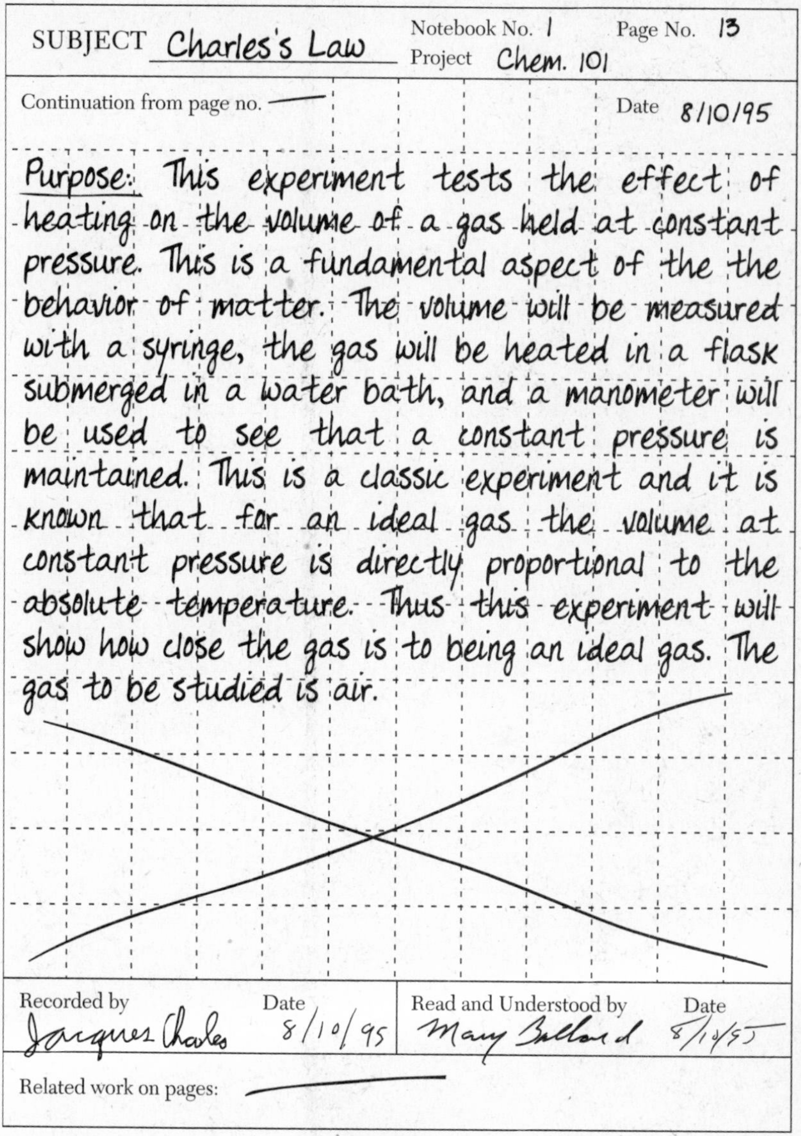

SUBJECT Charles's Law | Notebook No. 1 | Page No. 13 | Project Chem. 101

Continuation from page no. ——— | Date 8/10/95

Purpose: This experiment tests the effect of heating on the volume of a gas held at constant pressure. This is a fundamental aspect of the the behavior of matter. The volume will be measured with a syringe, the gas will be heated in a flask submerged in a water bath, and a manometer will be used to see that a constant pressure is maintained. This is a classic experiment and it is known that for an ideal gas the volume at constant pressure is directly proportional to the absolute temperature. Thus this experiment will show how close the gas is to being an ideal gas. The gas to be studied is air.

Recorded by Jacques Charles | Date 8/10/95 | Read and Understood by Mary Ballard | Date 8/10/95

Related work on pages: ———

FIGURE 4.1

Figure 4.2. Notice that in this kind of experiment a drawing of the apparatus is very important and takes the place of many words.

Procedures

The Procedures section of your notebook is written as the lab work progresses and records exactly what you did in the lab. You can judge the quality of your Procedures by asking yourself, "Will I be able to understand and repeat this experiment if I come back to this notebook in a couple of years?" and "Will another scientist be able to repeat this work exactly based on these notes?" Remember that every detail of your work may be crucial in reproducing your results. Don't leave out a thing. Do not wait until after the lab is over to write your notes. Your notebook should have its own place on the laboratory bench so you can write as you go.

Write in the first person and in the active voice, the way you speak and think. Despite what many students suppose, the first person and the active voice have an important role in scientific writing because together they describe who is acting and what they are doing in the laboratory. For example, say, "I heated 100 mL of distilled water to boiling in a 400 mL beaker" instead of, "The quantity of 100 mL of water having been introduced into a 400 mL beaker was caused to boil as the result of the application of heat." Not only is the first person/active voice example shorter and therefore easier and quicker to write, it is also more informative because it identifies who is doing what.

Notice in the sample procedure in Figure 4.3 that the writer mentions things that did not go as planned; the syringe was really hard to get into the stopper, the plunger in the syringe did not move freely as the temperature was raised, and he could not get all the water out of the flask when measuring its volume. These observations will be extremely important if the experiment is repeated at a later date or if improvements in the procedure are to be made.

As you are reading this section ask yourself, "Could I repeat this experiment using only the procedures given here?"

Results

The Results section is where you record all of your observations and data. This is likely to be the longest section that you will write on the project. In recording data you should describe what you observe *as you observe it.* If you notice anything that you don't expect, write that down also. Later on, you will need to account for any surprises or discrepancies that you encounter.

SUBJECT Charles's Law | Notebook No. 1 | Page No. 14 | Project Chem. 101

Continuation from page no. 13 Date 8/10/95

Plan: The apparatus for this experiment is:

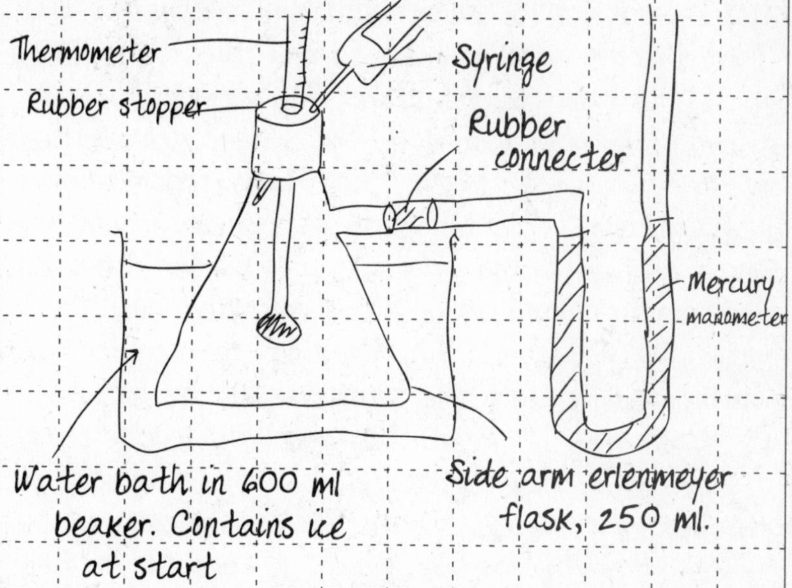

The flask is clamped on a ringstand so that it is suspended in the water bath. Drive the syringe needle through the stopper containing the thermometer. Place stopper in flask. Set syringe to 0.0 and then Attach to needle.

Recorded by Jacques Charles Date 8/10/95 | Read and Understood by Mary Ballard Date 8/10/95

Related work on pages: 13

FIGURE 4.2

SUBJECT Charles's Law	Notebook No. 1	Page No. 15
	Project Chem. 101	
Continuation from page no. 14		Date 8/10/95

Plan (continued). The flask of air will be cooled to 0.0°C with an ice bath with the syringe at 0.0. Then the ice bath is removed and the flask slowly warms to room temperature. The temperature of the air and the volume in the syringe are plotted every 5°C or so. When the flask reaches room temperature, a water bath is placed around it, which is heated with a bunsen burner. Temperatures and volumes are recorded every 5°C until 60°C where the heating is terminated. During the heating the syringe should be adjusted if necessary to keep the mercury levels equal assuring a constant pressure. After the heating is finished the volume of the flask is measured by filling it with water, which is poured into a graduated cylinder. The total volume of the gas is the volume of the flask plus the volume in the syringe.

Recorded by	Date	Read and Understood by	Date
Jacques Charles	8/10/95	Mary Ballard	8/10/95

Related work on pages: 13, 14

FIGURE 4.2 *(Continued)*

SUBJECT Charles's Law | Notebook No. 1 | Page No. 16 | Project Chem. 101

Continuation from page no. 15 | Date 8/10/95

Procedures: I assembled the apparatus as in the drawing on p. 14. Pushing the syringe needle through the rubber stopper was the most difficult part and I bent the needle on the first try. I cooled the flask to 0°C and then let it warm up. The syringe did not seem to move freely but moved all right if I rotated ~~it~~ the plunger in the syringe from time to time. The level of the mercury did not change indicating that the pressure in the flask was remaining constant. When the flask reached 22°C I put a water bath around it and heated this with a bunsen flame up to 62°C. Throughout all this procedure I recorded the temperature in the flask and the volume in the syringe every 5°C or so. When I measured the volume of the flask I was not able to get every bit of water into the graduated cylinder. This would cause some ~~a~~ error in the volume.

Recorded by Jacques Charles | Date 8/10/95 | Read and Understood by Mary Ballard | Date 8/10/95

Related work on pages: 13–15

FIGURE 4.3

The initial data that you collect are called *raw data.* These are the actual numbers written down from a weighing, an instrument, a titration, or any other measurement. Record the raw data carefully and generally leave interpretation and drawing conclusions until later.

Don't let your ideas of the moment influence the way you record your data. The trend that you seem to be seeing may not be real. In most cases, the raw data will be subjected to calculations and interpretation. The calculations and interpretation may be wrong, so it is essential to save the raw data so that you can go back to it.

The raw data recorded in Figure 4.4 are the actual numbers read in the experiment before any calculations have been made. The column labels on the table state clearly that the temperature is in °C, the common scale on thermometers, and the volume recorded is read directly off the syringe in mL. Conclusions are not appropriate at this point and none have been drawn. The calculations to be made are then described, and the calculated points are given in a second table. Then, as in many cases, the results are plotted on a graph, which makes interpretation easier. A brief statement of the observed result is given at the end of the section.

Discussion

The Discussion section gives you a chance to evaluate the results of your laboratory work—to think about how and why you got the results you did. If your data turned out as you expected, why is this the case? What principles of chemistry can explain the results? If your data turned out different than you expected, why is this the case? How would you explain it? The Discussion section uses writing as a tool to help you begin formulating interpretations and explanations.

In the sample discussion section in Figure 4.5, the writer immediately notes that the points on the graph do not give a perfectly straight line. The writer then explores the question of whether this is because the gas is not ideal or because of errors in the experiment. All the possible sources of error were listed to help answer this question. This leads the writer to the conclusion that experimental errors are the problem. Often the things that don't go quite right in an experiment are more interesting than the things that do.

Conclusions

The Conclusions section will normally be fairly short. It should summarize the goal of the work and evaluate how effectively that goal was reached and if anything unexpected was learned. It is natural and ap-

SUBJECT Charles's Law | Notebook No. 1 | Page No. 17 | Project Chem. 101

Continuation from page no. 16 | Date 8/10/95

Results: The temperatures and syringe volumes recorded were

Temperature, °C	Volume in Syringe, mL
0.0	0.0
5.2	4.4
9.9	9.0
14.7	13.1
22.0	20.5
29.7	26.0
36.1	31.9
41.0	37.4
45.5	40.9
51.0	47.5
56.3	50.9

Volume of flask, 248.2 mL.

This raw data was then converted of kelvin temperatures (by adding 273.15 to °C) and to total volume (by adding 248.2 mL to volume in syringe.)

See next page.

Recorded by Jacques Charles | Date 8/10/95 | Read and Understood by Mary Ballard | Date 8/10/95

Related work on pages: 13–16

FIGURE 4.4

SUBJECT Charles's Law | Notebook No. 1 | Page No. 18 | Project Chem. 101

Continuation from page no. 17 Date 8/10/95

Temperature, K	Volume of gas, mL
273.15	248.2
278.35	252.6
283.05	257.2
287.85	261.3
295.15	268.7
302.85	274.2
309.25	280.1
314.15	285.6
318.65	289.1
324.15	295.7
329.45	299.1

These values can be plotted to see if Charles's Law is obeyed. If it is I will see a straight line.

Recorded by Jacques Charles Date 8/10/95 | Read and Understood by Mary Ballard Date 8/10/95

Related work on pages: 13–17

FIGURE 4.4 *(Continued)*

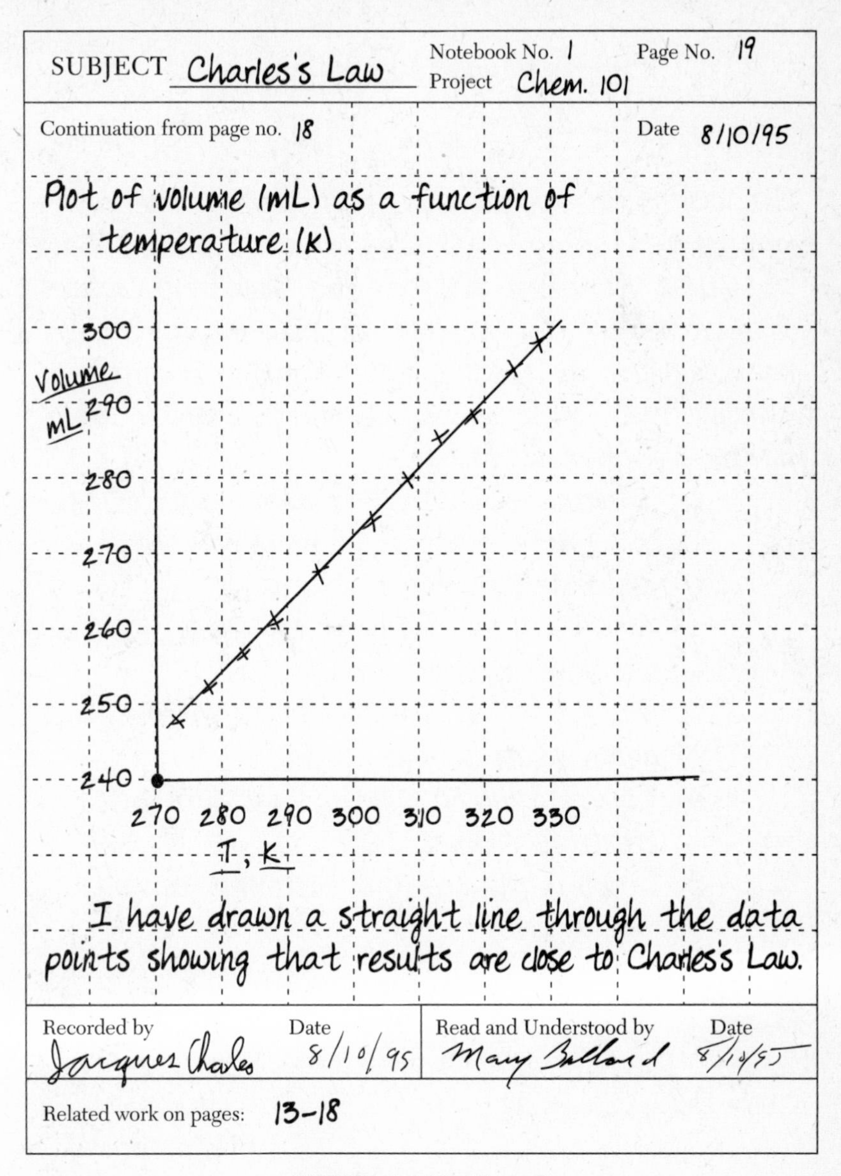
SUBJECT Charles's Law
Notebook No. 1
Page No. 19
Project Chem. 101
Continuation from page no. 18
Date 8/10/95
Plot of volume (mL) as a function of temperature (K)
Volume mL
300
290
280
270
260
250
240
270 280 290 300 310 320 330
T, K
I have drawn a straight line through the data points showing that results are close to Charles's Law.
Recorded by
Jacques Charles
Date 8/10/95
Read and Understood by
Mary Ballard
Date 8/10/95
Related work on pages: 13–18

FIGURE 4.4 *(Continued)*

SUBJECT Charles's Law

Notebook No. 1 Page No. 20

Project Chem. 101

Continuation from page no. 19 Date 8/10/95

Discussion: The plot on page 19 gave a line that was almost straight showing that Charles's Law was close to being obeyed. The slight deviations from a straight line could be because air is not behaving like an ideal gas or because of errors in the method. These are a number of possible errors in the procedure as follows:

1. Incorrect temperature as a result of incomplete transfer of heat to the thermometer bulb.
2. Air leakage during heating giving gas volumes that are too low.
3. Sticking of the plunger in the syringe again giving low gas volumes.
4. Failure to keep constant pressure giving erroneous volumes.
5. Errors in reading the thermometer and the syringe.

Recorded by Jacques Charles Date 8/10/95

Read and Understood by Mary Ballard Date 8/10/95

Related work on pages: 13–19

FIGURE 4.5

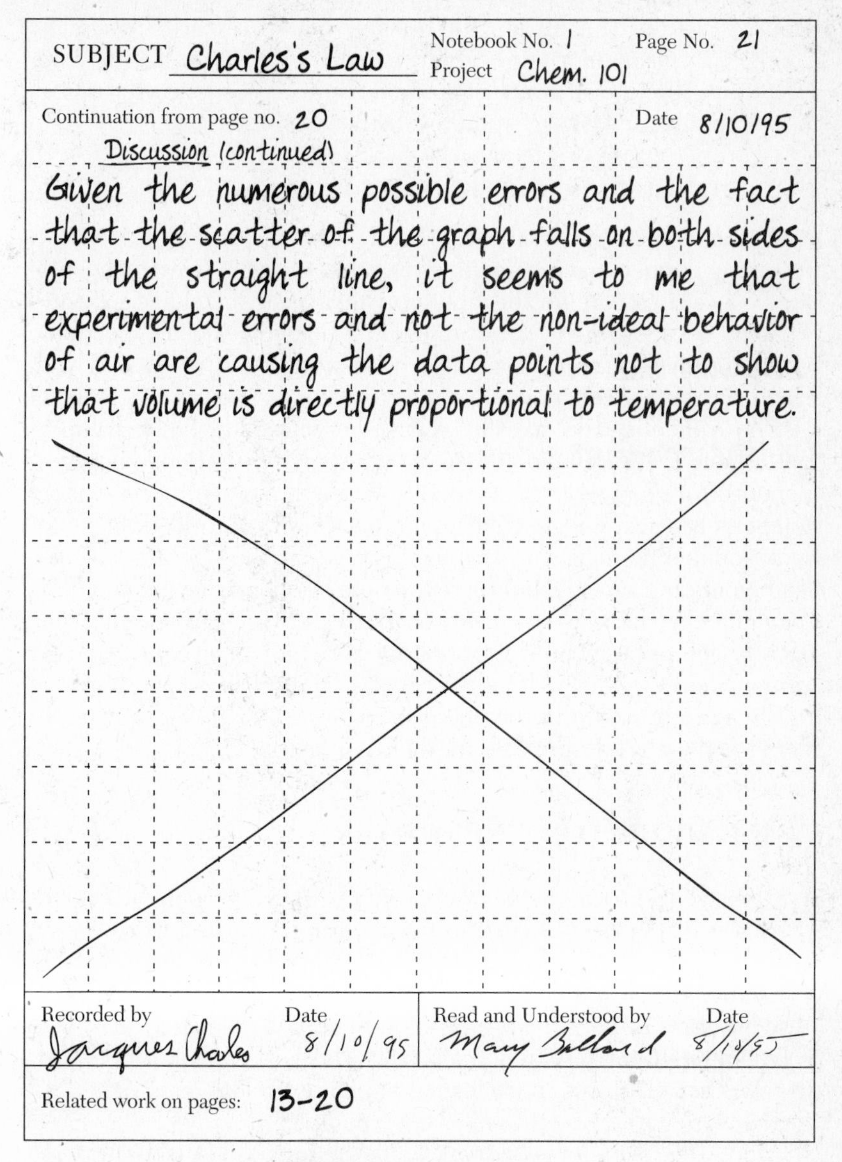

SUBJECT Charles's Law Notebook No. 1 Page No. 21
Project Chem. 101

Continuation from page no. 20 Date 8/10/95

Discussion (continued)

Given the numerous possible errors and the fact that the scatter of the graph falls on both sides of the straight line, it seems to me that experimental errors and not the non-ideal behavior of air are causing the data points not to show that volume is directly proportional to temperature.

Recorded by Jacques Charles Date 8/10/95
Read and Understood by Mary Ballard Date 8/10/95

Related work on pages: 13–20

FIGURE 4.5 *(Continued)*

propriate to put suggestions for further related research here. The sample conclusion in Figure 4.6 gives some of the writer's general observations, the overall conclusion made from the experiment, a brief interpretation, and a suggestion for further work.

Using the Computer for Keeping Records

Using the computer as a substitute for a written laboratory notebook is becoming popular and is sometimes referred to as the "electronic notebook." As a substitute for the laboratory notebook, the electronic notebook has advantages and disadvantages. The legibility of the electronic notebook, the ability to search by code-words, and the speed with which an experienced typist can enter data are all attractive. Storage of data on magnetic disks takes up very little space, although it is important to note that magnetic storage of data cannot be considered as permanent as a good written notebook. Although the written laboratory notebook is susceptible to changes and tampering, the electronic notebook is much more so. Instructions to "read and add but not change" can be put on a data file, but these are possible to circumvent and deletions and changes may occur. In recent cases of alleged scientific dishonesty, computer records of laboratory work have been considered extremely weak evidence. Although the use of the computer for collecting and filing information in the chemistry laboratory can be expected to increase, there is no substitute for the laboratory notebook.

THE LABORATORY REPORT

As you have just seen, the laboratory notebook is a sequential account of what was done in the laboratory. Its primary purpose is to keep a record of research in progress, so that nothing gets lost and either the researcher or co-workers can reconstruct experimental work. In a sense, it is an "in-house" working document, and its readers are likely to be limited to the researcher and co-workers. The laboratory report, on the other hand, is a more formal public document constructed after the work in the laboratory has been completed.

Too often chemistry students think the purpose of laboratory reports is to write up the results of a lab in order to prove to the chemistry instructor that they did the lab and got the expected results. While this view of the laboratory report may be understandable, it is unfortunate because it misses altogether the scientific purpose of the laboratory report—to explain why a researcher conducted a particular

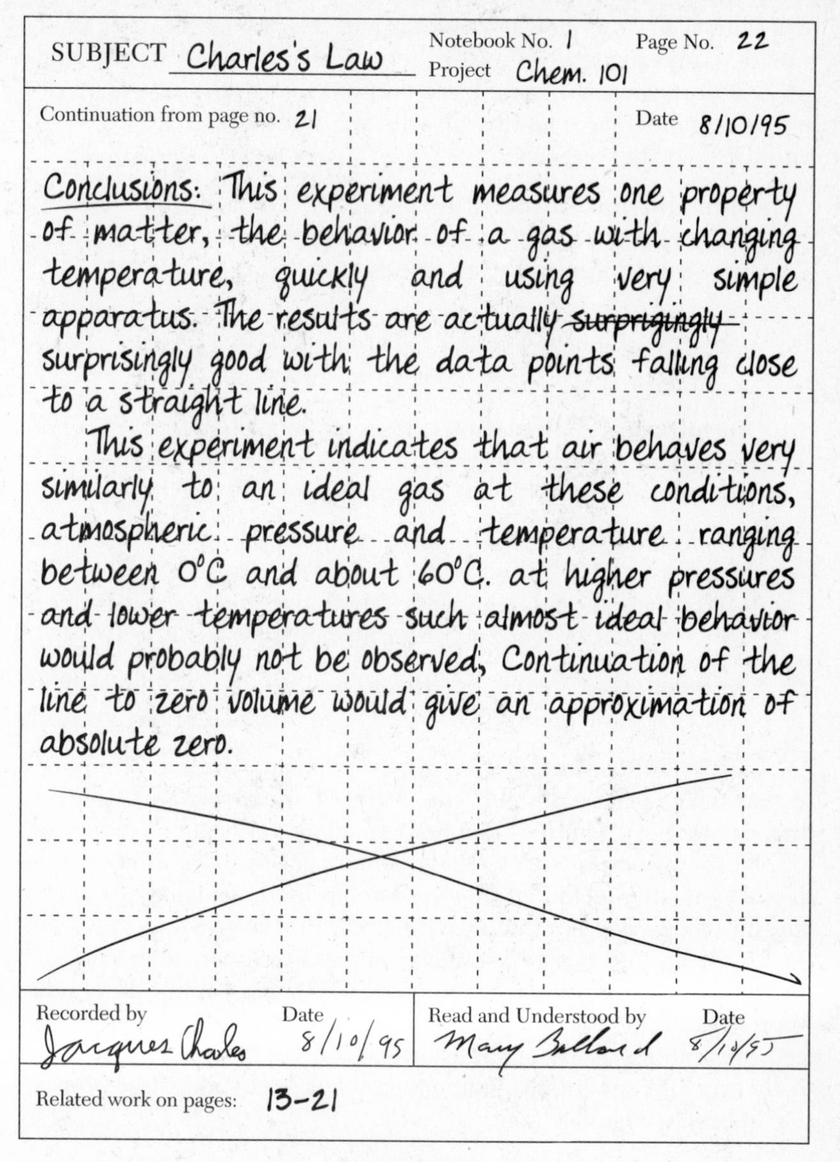

SUBJECT Charles's Law Notebook No. 1 Page No. 22
Project Chem. 101

Continuation from page no. 21 Date 8/10/95

Conclusions: This experiment measures one property of matter, the behavior of a gas with changing temperature, quickly and using very simple apparatus. The results are actually ~~surprigingly~~ surprisingly good with the data points falling close to a straight line.

This experiment indicates that air behaves very similarly to an ideal gas at these conditions, atmospheric pressure and temperature ranging between 0°C and about 60°C. at higher pressures and lower temperatures such almost ideal behavior would probably not be observed, Continuation of the line to zero volume would give an approximation of absolute zero.

Recorded by Jacques Charles Date 8/10/95
Read and Understood by Mary Ballard Date 8/10/95

Related work on pages: 13–21

FIGURE 4.6

experiment, to describe the methods and results, and to offer interpretations and suggestions for further research. In other words, laboratory reports have the same basic purpose that scientific articles do, and lab reports and scientific articles are organized in similar ways, with the following sections:

- **Introduction.** Explains the purpose of the research reported, using relevant background information to define the particular problem addressed by the research.
- **Procedures.** Describes how the research was done in enough detail to permit another researcher to perform the same experiment.
- **Results and Calculations.** Presents the actual data, observations, and measurements, as well as calculations made on this data; normally uses tables and graphs as well as written text.
- **Discussion and Conclusions.** Analyzes the results of the experiment, in relation both to the purposes stated in the "Introduction" and to available knowledge in the scientific literature. May offer suggestions for further study.
- **References.** Presents full and accurate citations for any references cited in the report.

Like scientific articles, laboratory reports are carefully composed so that readers can easily find and extract the information and ideas that are most relevant to them. Some readers, for example, will want to read the whole report or article. Some may be more interested in the experimental techniques the researcher used, while other readers may find the researcher's interpretations of the data more pertinent to their interests and purposes as chemists. Still another reader may be preparing a literature review and therefore be most interested in the references to previous work. In any case, as you can see, the formalized conventions of the scientific article and the laboratory report actually expedite communication among scientists by locating information in predictable places.

Writing the Laboratory Report

The following sections offer suggestions on writing laboratory reports. Make sure you allow yourself enough time to produce a lab report you can be proud of. To gauge the amount of time you will need, it will help to look at this overview of the writing process involved in producing a lab report:

1. Review your laboratory notes.
2. Plan and draft your report.
3. Design needed tables and graphs.
4. Revise the draft of your report.
5. Edit and proofread the final draft.

Suppose you have performed a laboratory experiment in which you were asked to use the ideal gas law to determine the molecular weight of carbon dioxide. The experimental technique is to vaporize a sample of dry ice, determine the mass (m), volume (V), temperature (T) and pressure (P) of the carbon dioxide vapor, and then calculate the molecular weight of the gas from the equation,

$$M = \frac{mRT}{PV}.$$

In this equation R is the universal gas constant, 0.0821 L • atm/mol • K.

Review Your Laboratory Notes

Before you begin planning and drafting your lab report, spend some time reading and re-reading your lab notes. How did you define the purpose of the experiment before you began? Do you have a full and accurate account of lab procedures? Are the results those you expected? What surprises or discrepancies, if any, did you encounter? Did you do any writing that evaluates your data or offers interpretations? Do you have the references you will need for your report?

Plan and Draft Your Report

Each section of a lab report or scientific article has its own particular function. Understanding these functions can help you to organize your material. Scientists usually write the Introduction last—because only after drafting the other sections do they know what they are introducing. In the discussion below, however, we start from the beginning and describe the function of each section in order.

Introduction

The initial part of the Introduction should explain the purpose of the experiment, beginning with a brief statement of why the experiment was performed and what problem or question it addresses. Then the strategy of the work to be done can be explained, followed, if appropriate, by the results of previous work in this area, with references.

In the initial statement of purpose, do not state that you plan to prove the truth of something. Rather, plan to test a hypothesis. This strategy of formulating and testing a hypothesis is at the very heart of the scientific method of inquiry. It may help to see that hypotheses, which researchers formulate to guide their experiments, can never be proven once and for all. Rather, hypotheses are discredited, modified, or supported, depending on experimental results. Hypotheses that are supported regularly by experimental evidence work their way into the scientific literature as established theories, but, in general, no theory can be considered final and definitive. All are subject to experimental confirmation or change.

Consider these three openings of the Introduction:

EXAMPLE ONE

```
    The purpose of this experiment was to complete
Experiment 11 which is assigned for the course.
```

EXAMPLE TWO

```
    The purpose of this experiment was to prove
that the molecular weight of carbon dioxide is
really 44 as it is calculated from the atomic
weight table.
```

EXAMPLE THREE

```
    The purpose of this experiment was to see how
closely the molecular weight of a real gas can be
determined using ideal gas calculations.
```

The first example is clearly hopeless. It is like saying that the purpose of going to college is to fill a seat for four years and get a degree. The second example actually states a real purpose; however, that purpose is unreasonable. If the calculation for the molecular weight of carbon dioxide comes out to be 49.0, does that mean that the atomic weight table is in error?

In contrast, the third example states real objectives that can be achieved. It does not try to "prove" anything but rather to test the assumptions involved in a simple method of determining the molecular weight of a gas.

The Introduction next describes the general strategy of the experiment and includes the important equations that will be used. In addition, this section of the Introduction also discusses any experimental difficulties that must be overcome.

Finally, cite your textbook, laboratory manual, lecture notes, and outside reading to reference background information and to define important terms. Here is a suitable example of the part of the Introduction that follows the initial statement of purpose for the experiment:

The basis for this experiment is the ideal gas law, $PV=nRT$, the equation of state for an ideal gas. The number of moles n equals $\frac{m}{M}$ where m is the mass of the sample and M is the molecular weight. If we substitute this expression into the ideal gas law and rearrange we get $M=\frac{mRT}{PV}$. We can determine all of the quantities on the right-hand side of this equation by experiment.

However, it is not easy to determine the value of m, the mass of the gas, directly. The problem here is that it is not practical to weigh the flask when it is empty since it is filled with air when it is not filled with CO_2. The problem is solved in this experiment by weighing the flask filled with CO_2 at two temperatures, T_1 and T_2, where T_1 is the lower. The method is described in detail in the lab book, *Laboratory Experiments for General Chemistry,* H. R. Hunt and T. F. Block, Saunders College Publishing, Philadelphia, 1990. The molecular weight is calculated from the equation,

$$M = \frac{\Delta m R T_2 T_1}{PV(T_2 - T_1)}$$

where Δm is the difference in the mass of the CO_2 at the two temperatures.

Finally, it is appropriate to summarize the experimental method that will be used. For this experiment, the following is an effective summary:

A flask is filled with carbon dioxide by allowing a piece of dry ice to sublime thus pushing all of the air and excess carbon dioxide out of a needle placed through a septum cap. The needle is removed after the dry ice has all vaporized and the flask is weighed at room temperature, T_1. The needle is then replaced in the flask and it is heated to temperature T_2 and weighed again.

Procedures

The Procedures section should not simply be a recitation of the instructions you have been given. Make sure that you write the Procedures in your own words so that anyone reading it would really be convinced that you understand what you have done. References to previous work by other investigators should be included in this section as appropriate.

Selecting the appropriate level and amount of detail is a principal problem. Consider this opening of the Procedures section in the lab to determine the molecular weight of carbon dioxide using the ideal gas law:

I broke up a large piece of dry ice by striking it with a hammer. Pieces of all sizes were formed and I picked one out. I took this piece, brushed the frost off it with a small piece of paper towel, and then dropped it into an erlenmeyer flask. I then pushed a syringe needle through the septum cap on the top of the erlenmeyer and waited. Then I took the needle out and

```
took the flask to the balance room where I
weighed it. . . .
```

Now think about what information is really necessary and what is missing in this opening. Let's consider the unnecessary details first. The size of the initial piece of dry ice from which the smaller pieces were broken is not important. And the fact that pieces of dry ice of many different sizes were formed has nothing to do with the experiment being performed, although it could be significant in some other experiment. What *is* important is the size of the sample that is put into the flask. What tool was used to break the dry ice is of marginal importance, but mentioning it is not inappropriate. This section also contains some unnecessary words such as, "I took this piece . . .," referring to the sample of dry ice used for the experiment.

A number of very important points are missing, and these missing points will give the reader the feeling that the writer was not exactly sure of what he or she was doing. For example, the size of the erlenmeyer flask is not given. Was it necessary for the erlenmeyer to be dry? Nowhere does it say that a septum cap was placed on the erlenmeyer flask, although this can be inferred from the fact that the syringe needle was pushed through a septum cap. A significant point is that when a septum cap is placed on a flask containing a piece of dry ice, the syringe needle had better be pushed in pretty quickly. Otherwise, the pressure of the CO_2 will build up and explode the flask. And the writer describes waiting before weighing the flask. Waiting for what or for how long? The waiting is for all of the solid to turn to gas so that the ideal gas law can be applied. Finally, the details of the weighing were not described. What kind of balance was used and to what precision was the weighing performed?

Here is a rewriting of the same section taking into account what should be deleted and what should be added.

I brushed the frost off a piece of dry ice about the size of a pea with paper towel and put it into a clean and dry 125 mL erlenmeyer flask. I put a septum cap over the mouth of the flask and *immediately* pushed a syringe needle through this cap to prevent excessive buildup of CO_2 gas. I took out the needle when all of the dry ice had

```
disappeared and quickly weighed the flask on an
analytical balance to the nearest 0.001 g. The
flask was handled with a piece of paper towel dur-
ing all operations to prevent getting fingerprints
on the glass, which would result in an erroneous
weight.
```

This revision is a little longer than the first but the amount of information and the evidence of understanding are much greater.

Results and Calculations

The Results and Calculations section presents in an orderly and readable form the measurements and specific observations originally recorded in the laboratory notebook. Be sure to include the units on every quantity that you measured or calculated. Also, be sure that you recognize the uncertainties in your measurements (see the next section, Uncertainties in Measurements) and that your final calculations have the correct number of significant figures.

How many digits are significant will vary from one experiment to another. In the case of the discovery of argon, Lord Rayleigh observed that the masses of equal volumes of nitrogen gas prepared from decomposition of ammonium nitrate and from the removal of all of the oxygen in air were different by a factor of 1.0053. Now, 1.0053 is very close to 1.0000, and he could simply have explained the difference as experimental error. Repeated measurement, however, showed the extra 0.0053 was, in fact, significant. By persistently exploring the discrepancy between the expected result of 1.0000 and the actual experimental result of 1.0053, Lord Rayleigh discovered that air contains 0.93% of a new element, the noble gas argon.

Here are one student's observations made on the determination of the molecular weight of carbon dioxide:

```
    The experiment worked pretty much as expected.
It was kind of a long wait before all of the dry
ice vaporized.
```

This explains little about what was actually observed, and the fact that the dry ice vaporizes rather slowly is not very significant unless it is coupled with a further observation that might explain its meaning or importance. A more satisfactory statement of observations is the following example:

```
    When the dry ice had vaporized, there was the
tiniest droplet of a liquid remaining. This was
probably water from frost that was on the piece of
dry ice because of condensation from the air or
failure to brush all the frost off. A water
droplet will cause an error since it will con-
tribute to the mass of the flask plus the CO2 gas.
We also noticed that the mass that we observed on
the analytical balance for the higher temperature
determination was increasing slowly, which was
probably the result of air leaking into the flask
through the septum as the temperature dropped.
```

Uncertainties in Measurements

In order to make the necessary calculations for completing an experiment, it is necessary to include the **uncertainties** or the **errors** in the measurements and in the results of calculations. Uncertainties and errors are not exactly the same thing. The uncertainty is a number that is estimated from the way that a measurement is made or from the results of a number of measurements of the same value. For example, in measuring the length of an object with a ruler where tenths of a millimeter must be estimated, the uncertainty is approximately 0.1 or 0.2 mm.

The error in a measurement is the difference between the measured value and the true value. The problem is that we normally don't know the true value. If we did, we wouldn't have to make the measurement. Because of the difficulty in actually determining the error, we usually make the assumption that the error and the uncertainty are fairly close in value and we use these two terms essentially interchangeably.

The techniques for dealing with errors in laboratory measurements are covered in textbooks and most lab books in general chemistry. Look up the subject of errors in your own text and be sure you understand the following concepts:

- Absolute error, relative error, and percent error;

- Significant figures;
- Rounding off; and
- What happens to uncertainties when measurements are added, subtracted, multiplied, divided, or appear in expressions involving logarithms.

Preparation of Tables

A table is a collection of information organized in rows (horizontal) and columns (up and down). Because it is a solid block of information, a table is useful for organizing a large amount of data and making it easier to read and understand. It is important to construct tables for yourself to organize your own work, and it is often necessary to make tables for reports, articles, and presentations. There are a number of guidelines for making tables that you should follow.

- A table should have a brief title, which should make the table understood without referring to any text.
- A table should have at least two columns.
- Like objects should be listed vertically, not horizontally.
- Numbers in a column should have their decimal points one directly above the other.
- Every column should be labeled with a heading giving the appropriate information. This heading normally has a maximum of two lines. The quantity and its units can be on separate lines or separated by a comma. Never head a column with just the units; use *Pressure, atm*, not just *atm*.

Our example of a table comes from repeating Boyle's experiment on the effect of pressure on the volume of a gas. The raw data from this experiment is the length of an air sample trapped in a J-shaped glass tube, which will yield the gas volume, and the difference in height of two columns of mercury, which will yield the pressure. Suppose for five determinations you measure lengths of the air sample of 122.5, 116.2, 111.4, 106.2 and 103.6 mm with an estimated uncertainty of ±0.2 mm. And the differences in the mercury columns for these five determinations are 51.0, 94.4, 131.7, 174.3, and 197.9 mm, also with an estimated uncertainty of ±0.2 mm. These numbers are your raw data and they can be organized in a table like the following for ease of interpretation and future reference.

Boyle's Law Experiment: Measuring Volume as a Function of Pressure, September 26, 1996

Trial #	*Length of air sample mm, ±0.2*	*Hg height difference mm, ±0.2*
1	122.5	51.0
2	116.2	94.4
3	111.4	126.7
4	106.2	174.3
5	103.6	197.9

Notice that the data are harder to read in a format that puts similar items in horizontal rows.

Trial	*1*	*2*	*3*	*4*	*5*
Length of air sample, mm, ±0.2	122.5	116.2	114.4	106.2	103.6
Hg height difference, mm, ±0.2	51.0	94.4	126.7	174.3	197.9

One more guideline in preparing tables is,

- The entry in the extreme left column normally refers to all entries in that row. Note in our table that the values in the pressure and volume columns both refer to the trial number in the left-hand column since these values were all measured in that particular trial.

Graphing Data

Graphs are particularly important for revealing the trends in your data and calculations. Many scientific discoveries have been made from studying graphs, and it is important to draw a good graph so that the information in it can be easily visualized. Furthermore, graphs are invaluable for presenting your results to other people and are often the most persuasive evidence for your point of view. A graph is particularly useful for determining the relationship between two variables, for example, the pressure and volume of a gas. The procedure normally involves plotting the individual data points, drawing a line through these points, and then often finding a mathematical equation that fits this line.

When the measurements in the table of raw data above were converted to pressure and volume, the following values were obtained.

Trial	***Volume of air mm³, ±300***	***Pressure torr, ±0.2***
1	6200	810.3
2	5800	849.7
3	5600	882.0
4	5300	929.6
5	5200	953.2

From the table above it is clear that as the pressure exerted on the gas increases, the volume decreases. However, in order to get a more meaningful picture of how the pressure and volume are related it is necessary to prepare a graph (Figure 4.7). Guidelines for plotting a graph are:

- Use graph paper for convenience, accuracy, and clarity. It can be written on directly or tracing paper can be used.
- Plot the independent variable along the X-axis (the horizontal axis) and the dependent variable along the Y-axis (the vertical axis). The independent variable is the variable that you are changing directly, in this case, the pressure, which you are changing by increasing the length of the mercury column. The dependent variable is the variable that is responding to the change in the independent variable. In this case the dependent variable is the volume, which is changing in response to the change of pressure.
- The lower left-hand corner of the graph does not have to be the origin where independent and dependent variables both equal zero. It is often convenient to exclude the origin on a graph since it usually results in much wasted space. Ideally, you should label the axes of your graph so that most of the paper is

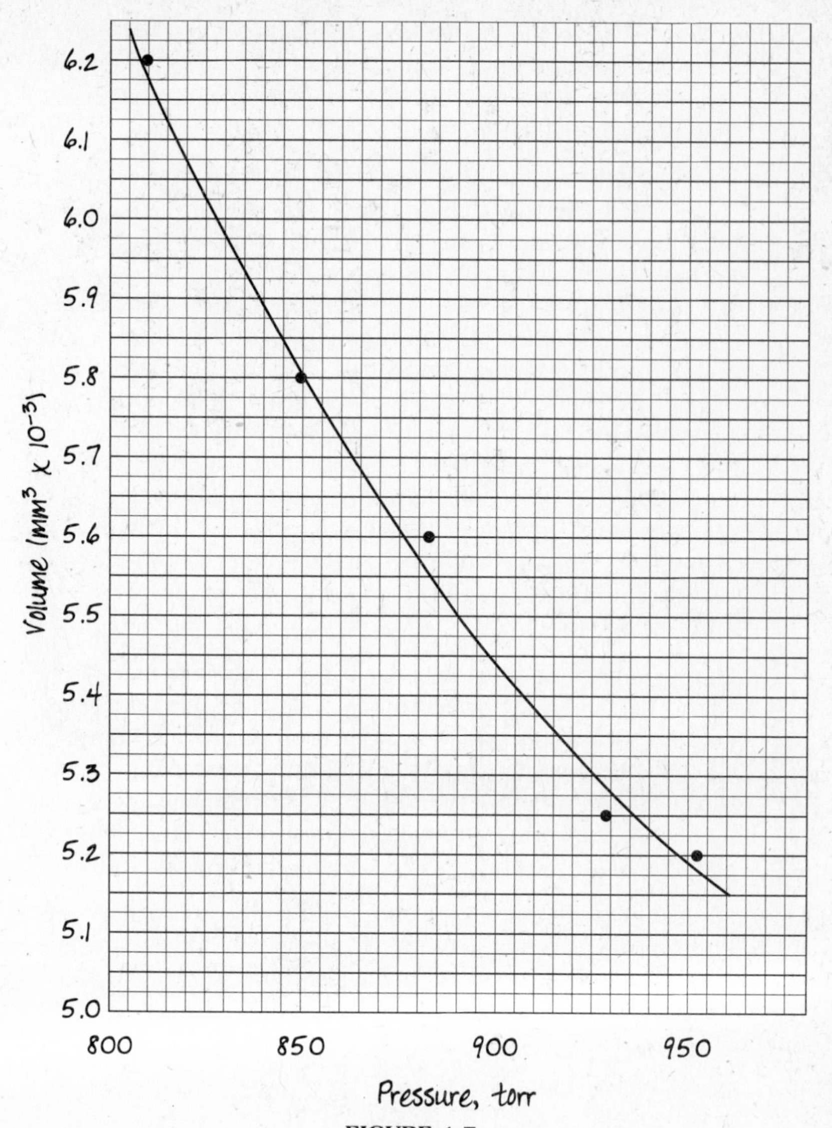

FIGURE 4.7

Boyle's Law Experiment, Volume as a Function of Pressure

utilized. That is, the range along the X-axis should only be a little larger than the range of the values of the independent variable and the range along the Y-axis should only be a little larger than the range of the values of the dependent variable.

- Use shorthand if the numbers along an axis are very large or very small. Otherwise these numbers will clutter the page. Such numbers along an axis are brought into a reasonable range by multiplying them by a constant factor. The simplified numbers are placed along the axes. *The number that the actual values have been multiplied by* is written either at the left side of the graph (Y-axis) or the bottom (X-axis). In our example, the range of values along the Y-axis is 5000 – 6200 and these values can be simplified to 5.0 – 6.2 by multiplying by 0.001 or 10^{-3}. Therefore the Y-axis values in Figure 4.7 are in the range of 5.0 – 6.2 with a notation that the values have been multiplied by 10^{-3}. These values must be *divided* by 10^{-3}, which is the equivalent of multiplying by 10^3. Thus if you read the volume of 5.8 off the graph you must divide this by 10^{-3} to get the actual volume of 5.8×10^3 mm^3 or 5800 mm^3.

- Draw a smooth line that gets as close as possible to at least most of the data points. Such a line will be most likely to yield a mathematical equation if desired. A French curve is very useful for drawing a smooth curved line. Simply "connecting the dots" is usually not a useful exercise. Scientists normally assume that there is at least some error in the data points and so do not expect that all of the points will lie on the line that you draw.

A straight line is the most desirable result to obtain in a graph since it can be fit most easily to a mathematical equation. Often changing a parameter or taking its reciprocal or its logarithm will yield a straight line. For example, a plot of volume (Y-axis) as a function of pressure (X-axis) actually gives a curved line called a hyperbola, whereas plotting the reciprocal of the volume $\left(\frac{1}{\text{volume}}\right)$ against pressure should yield a straight line that goes through the origin (Figure 4.8). In this graph the numbers along the Y-axis had to be multiplied by 10^4 or 10,000 to get a reasonable magnitude. To convert these back to mm^{-3} we must divide the values on the graph by 10^4. For example, 1.8 on the graph becomes $\frac{1.8}{10^4}$, which is 1.8×10^{-4} mm^3 or 0.00018 mm^3.

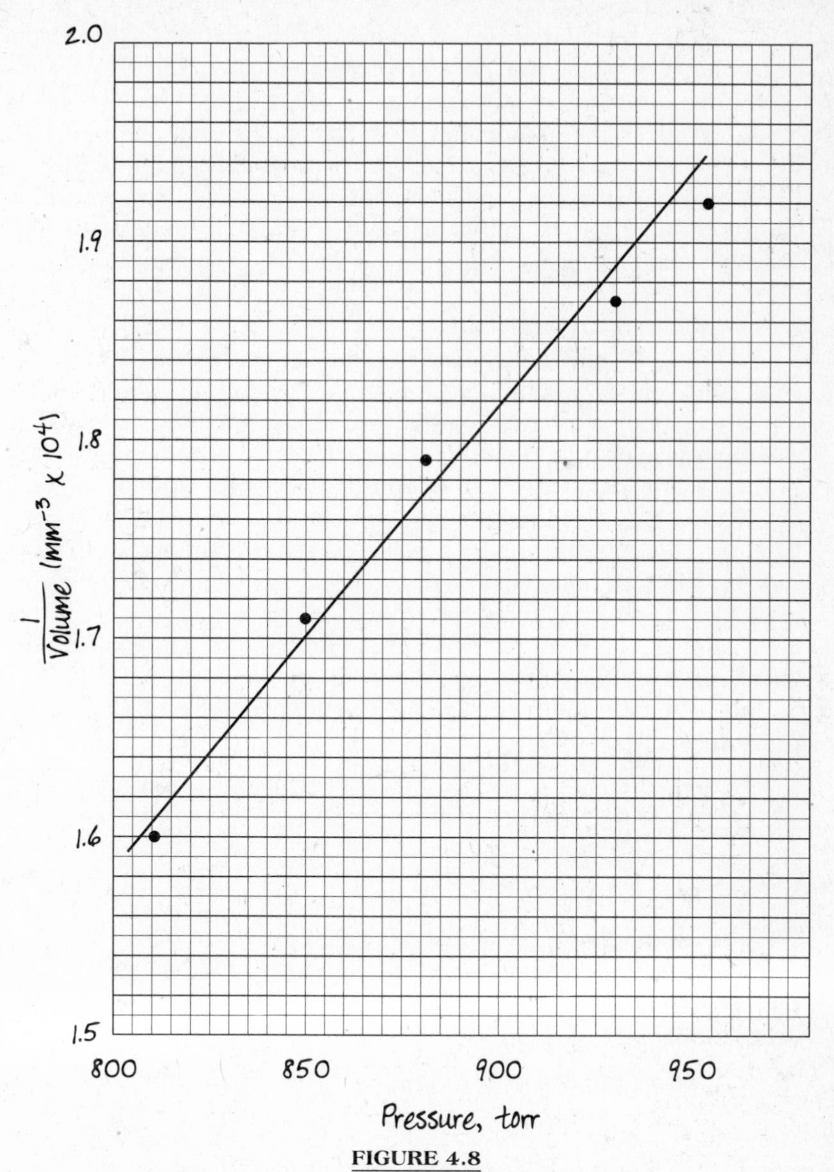

FIGURE 4.8

Boyle's Law Experiment, 1/Volume as a Function of Pressure

Graphing by hand as described above is becoming less common as computer graphing programs are used more and more. The use of such programs is often only a matter of entering the data. The program will select the coordinates, plot the points, draw the best line through them, and give the mathematical equation of this line in terms of the dependent and independent variables. However, it is worth learning how to graph manually in order to understand the strengths and limitations of graphing as a means of interpreting data.

Discussion and Conclusions

Your Discussion and Conclusions section is important to the overall success of your report because it is in this section that you examine the observations you have recorded and explain what they mean. You should usually discuss the effects of experimental error on the results in this section. State any assumptions that your conclusions depend on. Remember that practically any conclusion you make using data from the chemistry laboratory is based on some assumptions. Finally, sum up the experiment. How did it work and why? What did you learn?

Consider the two following examples of the Discussion and Conclusions section of the experiment above, duplicating Boyle's experiments to determine the dependence of the volume of a gas on its pressure:

EXAMPLE 1

I determined the volume of the gas as a function of pressure and plotted this on a graph. The graph looked curved so we tried plotting $\frac{1}{\text{volume}}$ as a function of pressure and got a straight line. Therefore we concluded that Boyle's law, which says that pressure times volume is a constant, is correct.

EXAMPLE 2

I calculated the volume of the air sample from the raw data as shown in the Results and Calculations section of this lab report. I noticed that the relative uncertainties in the pressures were quite small whereas the relative uncertainties in the volumes were much larger. The main reason for this was that calculating the volumes included the uncertainty in the radius of the glass tubing, which was squared in the calculation. Determining the pressure simply required measuring the differences in the mercury levels and the atmospheric pressure and both of these could be determined quite precisely.

The uncertainty in the volumes made the graph harder to interpret. Furthermore, it was hard to tell the shape of the curve because the region of pressures and volumes was relatively small since it is not practical to have a column of mercury that is very long. The plot of volume as a function of pressure looked a lot like a straight line. We really needed to know that it couldn't be straight ahead of time. The plot of $\frac{1}{\text{volume}}$ as a function of pressure looked more like a straight line and appeared to show agreement with Boyle's law that pressure times volume is a constant for air at the observed pressures.

The deviation that we observed from the straight line probably wasn't because the gas was not behaving ideally since the points appeared to be scattered on either side in a random manner. Determining more data points would help show if the error was just random. The reason for random error is probably the large uncertainty in the volume of the air as a result of the way that it was measured.

Notice that Example 1 simply repeats what the writer did in the laboratory and, without offering any explanation, concludes that Boyle's law is "correct." This discussion and its conclusion may give the impression that the student wrote it before or without doing and analyzing the actual laboratory work. Example 2, on the other hand, provides a more complete and thoughtful discussion of the experimental results, with explanations to account for the errors in the calculations of pressure and volume. Then instead of claiming, as Example 1 does, that the results prove that Boyle's Law is "correct," Example 2 compares the experimental results and calculations to what might be predicted according to Boyle's Law and then attempts to account for the deviations observed. Clearly, Example 2 reveals a student thinking like a chemist.

Different laboratory instructors will require different kinds of laboratory reports including previously prepared sheets to enter data and conclusions. However, the suggestions given above should be useful in practically all cases.

Revision

Once you have a draft of your laboratory report completed, take the time to read it carefully and to determine what changes, clarifications, additions, or deletions you want to make before you turn it in to your instructor. It can be helpful at this point to ask a classmate to read your report as well. Here are some questions to guide reading and revising:

- Is your material organized clearly into the conventional sections—Introduction, Procedures, Results and Calculations, and Discussion and Conclusions? Re-read the section above that describes the function of each section, and use these descriptions as guidelines to make sure your writing is organized properly.
- Do the sections fit together in a coherent way that explains what you did, why you did it, how you did it, what your results are, and what your results mean? Is the purpose of the experiment clearly stated in the Introduction? Does the Procedures section explain how you addressed the questions raised in the Introduction experimentally? Do the Results and Calculations appear as possible answers to the questions raised? Does the Discussion and Conclusion section clearly explain your results and what they indicate about the purpose of the experiment?
- Are there any sentences or passages in your draft that are confusing or hard to follow? Underline (or ask a classmate to underline) these sentences or passages so that you can go back and revise them.

- Are the tables and graphs clearly drawn and easy to follow? Are they properly labeled?

Editing and Proofreading

Make sure you re-read your report carefully before you turn it in to detect misspellings, grammatical errors, awkward sentences, and so on. Do this after you have revised your draft and made any substantive changes in the meaning and content of your report.

5

How to Read a Scientific Article: Writing Summaries and Critiques

The articles scientists publish in specialized journals are key means of communication within the various fields of study that make up the larger scientific community. While it is true that scientists communicate in a variety of other ways, including oral presentations at seminars, professional meetings, and conferences; phone calls; and email, the published article is perhaps the most important vehicle scientists use to make their work public—to present their research, describe their findings, and explain the significance of their work. By publishing the results of their work, scientists seek to participate in the ongoing conversation that defines the various fields of science, to establish the validity and significance of their research, and to secure their own position and reputation within a particular field.

For students learning science, reading articles that appear in the scientific journals is important because it brings them into contact with how scientists construct and validate new and reliable knowledge. Unlike, say, a chemistry textbook, which largely reports knowledge that is already agreed on by the community of chemists, articles that appear in scientific journals must make their claims persuasive to readers. Reading scientific articles, therefore, is not simply a matter of reading to acquire information. It also involves reading for the way scientists seek to convince other scientists that their research is valid.

In this chapter, you will see in greater detail how scientific articles try to establish persuasive claims, and we will offer some strategies for reading the literature of science. We will also look at two of the most common forms of writing assigned in response to reading scientific articles—summary and critique.

UNDERSTANDING SCIENTIFIC ARTICLES

Scientific articles are highly conventionalized forms of writing. They are compact, use specialized vocabulary, and generally follow a predictable pattern of organization. The compactness and specialized vocabulary of scientific articles often make them intimidating to newcomers to a particular field of study. On the other hand, the predictable pattern of organization can offer readers a strategy for following the line of argument in an article, even if they do not know all the scientific terms used in the article. In fact, as you will see, readers who do not necessarily have a thorough knowledge of the science described in an article can still analyze the persuasive techniques scientists use and thereby gain an understanding of what a particular scientist or group of scientists is trying to accomplish in an article. In this sense, learning to read scientific articles, even in those cases where you do not fully grasp all the scientific terms, concepts, and methods, can help you understand how scientists seek to establish the validity of their claims.

As mentioned in Chapter 4, most scientific articles use a form of organization similar to that used in laboratory reports, regardless of whether the article actually gives a heading for each section. The usual pattern of organization is:

- An *introduction* that defines the problem addressed in the article and explains why it is significant.

- A *literature review* that may either be part of the introduction or a separate section. The purpose of the literature review is to show how the problem addressed in the article grows out of prior work.

- In experimental articles, a section on *procedures*, sometimes labeled Materials and Methods, which describes exactly how the research being reported was conducted.

- A section on *results* that describes the researcher's findings—whether they are experimental data or a theoretical model—often using tables, graphs, or other types of illustration.

- A section of *discussion* that draws conclusions from the data or the model, makes claims about its significance, and points toward further experimental work that might be undertaken or larger problems that might be addressed by researchers.
- A list of *references* that cites published material the article draws on.

Once you have a sense of how scientists organize their material in scientific articles, you can use this basic format of scientific writing to identify how scientists go about making arguments for the importance of their work. Keep in mind that scientific articles do not simply report what a researcher has done but, more important, they advance claims in order to establish their significance. Scientists often refer to this significance as the "story" they can tell based on their research. This is not exactly the same kind of story found in fiction. Rather it is a meaningful account of how particular experimental research contributes to the wider understanding among scientists of a particular problem or issue.

To make their "story" persuasive to other scientists, researchers typically need to do two things. First, they need to show how their work addresses a real and well-defined problem or issue that other scientists will recognize as important. Second, they need to show how the results of their work enable them to make claims that will contribute to the wider understanding of the problem or issue at hand. If you keep these two purposes in mind, you will see the format of the scientific article as something more than a conventionalized format for reporting results. You can read scientific articles for the way they locate researchers in relation to previous work and then seek to portray the researcher's results as significant contributions to the field of study.

READING SCIENTIFIC ARTICLES

To demonstrate how reading the "story" of scientific research can help you understand scientific articles, let's take a look at one of the most important publications in twentieth-century science, James D. Watson and Francis H. C. Crick's article "A Structure for Deoxyribose Nucelic Acid," reprinted from the British journal *Nature*, where it appeared in April, 1953. This is the article in which Watson and Crick first reported their model of the structure of DNA, for which they won the Nobel Prize.

As you read Watson and Crick's famous article, notice that we have *underlined* and *annotated* it. Underlining and annotating are useful techniques for reading scientific articles. These techniques have different purposes for the reader:

- *Underlining* is a good way for readers to establish for themselves the main points or the content of the article.
- *Annotating*, or writing in the margins, is a good way for readers to keep a record of what the article's writers are doing and how they are developing their argument to persuade other scientists.

A Structure for Deoxyribose Nucleic Acid

James D. Watson and Francis H. C. Crick

1 We wish to suggest a structure for the salt of deoxyribose nucleic acid (DNA). This structure has novel features which are of considerable biological interest.

Presents major claim of paper, structure has novel features

2 A structure for nucleic acid has already been proposed by Pauling and Corey.[1] They kindly made their manuscript avaliable to us in advance of publication. Their model consists of three intertwined chains, with the phosphates near the fibre axis, and the bases on the outside. In our opinion, this structure is unsatisfactory for two reasons: (1) We believe that the material which gives the X-ray diagrams is the salt, not the free acid. Without the acidic hydrogen atoms it is not clear what forces would hold the structure together, especially as the negatively charged phosphates near the axis will repeal each other. (2) Some of the van der Waals distances appear to be too small.

Presents and refutes prior work by Pauling and Corey and by Fraser

[1]Pauling, L., and Corey, R. B., *Nature*, 171, 346 (1953): *Proc. U.S. Nat. Acad. Sci.*, 39, 84 (1953).

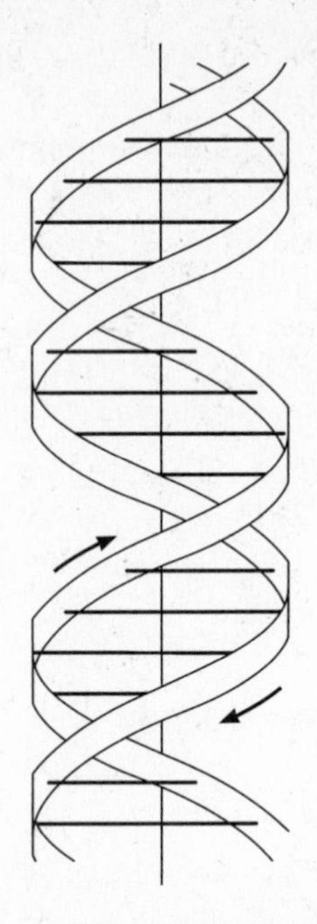

Visualizes the Structure

This figure is purely diagrammatic. The two ribbons symbolize the two phosphate-sugar chains, and the horizontal rods the pairs of bases holding the chains together. The vertical line marks the fibre axis.

3 Another three-chain structure has also been suggested by Fraser (in the press). In his model the phosphates are on the outside and the bases on the inside, linked together by hydrogen bonds. This structure as described is rather ill-defined, and for this reason we shall not comment on it.

4 We wish to put forward a radically different structure for the salt of deoxyribose nucleic acid. This structure has two helical chains each coiled round the same axis (see diagram). We have made the usual chemical assumptions, namely, that each chain consists of phosphate diester groups joining β-D-deoxyribofuranose residues with 3',5' linkages. The two chains (but not their bases) are related by a dyad perpendicular to the fibre axis. Both chains follow right-handed helices, but owing to the dyad the sequences of the atoms in the two chains run in op-

Describes structure (double helix) and assumptions

posite directions. Each chain loosely resembles Furberg's[2] model No. 1; that is, the bases are on the inside of the helix and the phosphates on the outside. The configuration of the sugar and the atoms near it is close to Furberg's "standard configuration," the sugar being roughly perpendicular to the attached base. There is a residue on each chain every 3-4 Å. in the z-direction. We have assumed an angle of 36° between adjacent residues in the same chain, so that the structure repeats after 10 residues on each chain, that is, after 34 Å. The distance of a phosphorus atom from the fibre axis is 10 Å. As the phosphates are on the outside, cations have easy access to them. 5

Gives prior credit

The structure is an open one, and its water content is rather high. At lower water contents we would expect the bases to tilt so that the structure could become more compact. 6

Describes and explains "novel feature"

The novel feature of the structure is the manner in which the two chains are held together by the purine and pyrimidine bases. The planes of the bases are perpendicular to the fibre axis. They are joined together in pairs, a single base from one chain being hydrogen-bonded to a single base from the other chain, so that the two lie side by side with identical z co-ordinates. One of the pair must be a purine and the other a pyrimidine for bonding to occur. The hydrogen bonds are made as follows: purine position 1 to pyrimidine position 1; purine position 6 to pyrimidine position 6. 7

If it is assumed that the bases only occur in the structure in the most plausible tautomeric forms (that is, with the keto rather than the enol configurations) it is found that only specific pairs of bases can bond together. These pairs are: adenine (purine) with thymine (pyrimidine), and guanine (purine) with cytosine (pyrimidine). 8

In other words, if an adenine forms one member of a pair, on either chain, then on these assumptions

[2]Furberg, S., *Acta Chem. Scand.*, 6, 634 (1952).

the other member must be thymine; similarly for guanine and cytosine. The sequence of bases on a single chain does not appear to be restricted in any way. However, if only specific pairs of bases can be formed, it follows that if the sequence of bases on one chain is given, then the sequence on the other chain is automatically determined.

Draws an inference

9 It has been found experimentally[3, 4] that the ratio of the amounts of adenine to thymine, and the ratio of guanine to cytosine, are always very close to unity for deoxyribose nucleic acid.

Reports compatible experimental evidence

10 It is probably impossible to build this structure with a ribose sugar in place of the deoxyribose, as the extra oxygen atom would make too close a van der Waals contact.

11 The previously published X-ray data[5,6] on deoxyribose nucleic acid are insufficient for a rigorous test of our structure. So far as we can tell, it is roughly compatible with the experimental data, but it must be regarded as unproved until it has been checked against more exact results. Some of these are given in the following communications. We were not aware of the details of the results presented there when we devised our structure, which rests mainly though not entirely on published experimental data and stereochemical arguments.

Notes limits of experimental evidence and calls for further work

12 It has not escaped our notice that the specific pairing we have postulated immediately suggests a possible copying mechanism for the genetic material.

Notes one major implication

13 Full details of the structure, including the conditions assumed in building it, together with a set of co-ordinates for the atoms, will be published elsewhere.

14 We are much indebted to Dr. Jerry Donohue for constant advice and criticism, especially on inter-

[3]Chargaff, E., for references see Zamenhof, S., Brawerman, G., and Chargaff, E., *Biochim. et Biophys. Acta*, 9, 402 (1952).

[4] Wyatt, G. R., *J. Gen. Physiol.*, 36, 201 (1952).

atomic distances. We have also been stimulated by a knowledge of the general nature of the unpublished experimental results and ideas of Dr. M. H. F. Wilkins, Dr. R. E. Franklin and their co-workers at King's College, London. One of us (J.D.W.) has been aided by a fellowship from the National Foundation for Infantile Paralysis.

Acknowledges and thanks co-workers and collaborators

Notice here that although Watson and Crick do not use the headings we have detailed above, their article still follows the general movement of a typical scientific paper. The following outline shows the various stages of Watson and Crick's argument:

Paragraphs 1–3: Introduction and Literature Review

First, Watson and Crick make clear the claim their paper is advancing: "We wish to suggest a structure . . . " that has "novel features which are of considerable biological interest." Readers of the *Nature* article will have noted right away how this paper represents other researchers' claims to have identified the structure of DNA, one of the central problems in biochemistry in the early 1950s. Moreover, they will note that Watson and Crick in the next two paragraphs review—and reject—two previously suggested models, namely Pauling and Corey's (in paragraph 2) and Fraser's (in paragraph 3). This sets up Watson and Crick to describe their own model.

Paragraphs 4–11: Results

In this section of the article, paragraphs 4 and 5 sketch the broad outlines of the model proposed. Paragraphs 6–8 detail the "novel features" that distinguish Watson and Crick's model. Paragraphs 9 and 11 consider the model in light of available experimental data, while paragraph 10 asserts that it is "probably impossible" to find the same model in RNA.

Paragraphs 12–14: Discussion and Conclusion

Paragraph 12 is perhaps one of the greatest understatements in the history of science. Watson and Crick say that it has "not escaped our notice" that the specific pairing in their model indicates the possibility of a "copying mechanism for the genetic material," exactly what other scientists have capitalized on since in molecular biology and genetic engineering. The final two paragraphs indicate that more information will be available in a forthcoming article and acknowledge the help of co-workers.

As you have seen, Watson and Crick's article in *Nature* does not contain a separate section on procedures or materials and methods. The reason for this is telling. Engaged in a race with other laboratories to identify the structure of DNA, Watson and Crick wanted to publish a model of DNA as soon as they felt confident of it, to establish their claim to have solved this scientific problem. The details, as they say, will follow.

WRITING SUMMARIES

Writing a summary of a scientific article is a useful next step after you have read, underlined, and annotated. Writing a summary demands that you understand the article well enough to describe it in your own words. In fact, putting what a scientific article says in your own words is a good way for you to grasp its meaning.

To help you prepare to write summaries, which are usually one to two typed pages, consider the following questions:

- What is the central claim of the article? Can you state this claim in one or two sentences of your own words?
- What problem or issue does this claim address? What clues can you find to how this problem or issue emerges from prior work in the field?
- What specific questions does the article address? Can you state each question in a separate sentence?
- How do the scientists address these questions? What specific approaches—theoretical, experimental, or both—are used to address the questions you've listed? What premises are these approaches based on?
- What are the major findings, whether experimental results or theoretical models? How does the article portray the significance of these findings?

- Are there questions that remain unanswered or new questions that emerge from the article?

Once you have considered these questions, you are ready to begin writing a summary. Here is a sample summary, based on Watson and Crick's article "A Structure for Deoxyribose Nucleic Acid":

> Watson and Crick propose a double helical structure for the salt of deoxyribose nucleic acid (DNA). After rejecting models of three intertwined chains by Pauling and Corey and by Fraser, Watson and Crick describe the "novel feature" of their structure, a mechanism by which two chains are held together. Each chain contains four bases that form a hydrogen bond by pairing a purine and a pyrimidine. According to the proposed structure, only specific pairs of bases can bond together. Adenine (a purine) pairs only with thymine (a pyrimidine), and guanine (a purine) pairs only with cytosine (a pyrimidine).
>
> While the sequence of the bases on either of the chains does not seem to follow any predictable pattern, the proposed structure makes it possible to predict the sequence of bases on one chain once the sequence of bases on the other chain is determined. Moreover, the proposed structure helps to explain the prior observation that the ratio of adenine to thymine and of guanine to cytosine is always one to one or very nearly. These ratios had been an unexplained curiosity, but in Watson and Crick's proposed structure, they become an inevitability.
>
> Watson and Crick suggest that previously published X-ray data appear to be compatible with their structure and the experimental results on which it is based but acknowledge that more work needs to be done before these X-ray data can constitute a reliable test of their proposed structure. Finally, Watson and Crick note that the specific pairing mechanisms in their proposed structure suggest the possibility of a mechanism to copy genetic material.

WRITING CRITIQUES

Writing a critique is similar in many respects to writing a summary. Like a summary, a critique identifies the central problem or issue, defines the central claim, looks at the specific questions, notes experimental and theoretical approaches, and reviews the results and their significance. What a critique adds to a summary is the writer's own analysis and evaluation of the article. This does not mean, however, that the writer should seek only to point out the faults or flaws in an article. A critique should emphasize first what the article contributes to the field and then identify shortcomings or limitations. In other words, a critique is a balanced appraisal, not a hatchet job.

To write a critique of a scientific article, begin by answering the questions on pages 98–99. Then take these questions into account:

- What is the strength of the article? What new information or method does it bring to light? How significant is the information or method?
- Do the article's conclusions seem justified by the data presented? Does the article generalize or make claims that are not supported by the data?
- Can you think of ways to improve the research described or to expand its scope? How would you address questions that remain unanswered?

Below we reprint an article and a sample critique.

Removal of Radionuclides in Wastewaters Utilizing Potassium Ferrate(VI)

Michael E. Potts, Duane R. Churchwell

ABSTRACT: *Jar tests were performed to demonstrate the applicability of using potassium ferrate(VI) [K_2FeO_4] to treat wastewaters containing americium and plutonium. Jar test treatment of tap water spiked with americium and plutonium with potassium ferrate lowered the gross alpha activity from 3.0 × 10^6 pCi/L to 6,000 pCi/L. An optimum treatment pH of 11.5 to 12.0 was found. At a Department of Energy (DOE) Facility demonstration, potassium ferrate treatment lowered gross alpha activity from 37,000 pCi/L to 40 pCi/L utilizing a two-step treatment process. The tests performed demonstrated that using potassium ferrate in the aqueous treatment plant at the DOE facility instead of the current treatment chemicals would ensure treatment requirements and discharge limits are met.* Water Environ. Res., ***66,*** *107 (1994).*

KEYWORDS: *coagulation, flocculation, potassium ferrate(VI), radioactive wastewater, solid–liquid separation*

Potassium ferrate has been evaluated as a multi-purpose wastewater treatment chemical for disinfection, oxidation, and coagulation (DeLuca *et al.*, 1983; Waite, 1981; Murmann and Robinson, 1974).

Research studies have indicated that potassium ferrate is also a more effective coagulant than other inorganic coagulants, such as aluminum, ferric, or ferrous salts (DeLuca *et al.*, 1992; Waite, 1981; Waite and Gray, 1984; Deininger, 1991).

Coagulation in water treatment by inorganic salts occurs primarily by adsorption of hydrolysis species on the colloidal particles resulting in charge neutralization, and the interaction between the colloid and the precipitating hydroxide resulting in sweep coagulation (Amartharajah and Tambo, 1991). The initial addition of aluminum or iron salts to wastewater results in the formation of soluble hydrolytic, polymeric species within microseconds after addition to the wastewater (Amartharajah and Tambo, 1991; Waite and Gray, 1984). The hydrolytic species are readily adsorbed at the liquid–solid interface of the colloid. The formation of the metal hydroxide precipitate occurs in the range of 1 to 7 seconds (Amartharajah and Tambo, 1991). The most important step in the coagulation process is charge neutralization to ensure that the hydrolytic products develop and destabilize the colloidal particles.

Potassium ferrate differs from other inorganic salts because the ferrate anion (FeO_4^{2-}) is soluble and disperses rapidly in aqueous solutions (approximately 15 g/L) (Schreyer and Ockerman, 1951). The ferrate anion (FeO_4^{2-}) decomposes rapidly in acid solutions (equation 1), but its stability increases with increasing solution pH (equation 2) (Schreyer and Ockerman, 1951).

$$FeO_4^{2-} + 8H^+ + 3e^- \rightarrow Fe^{3+} + 4H_2O \quad (1)$$

$$FeO_4^{2-} + 4H_2O + 3e^- \rightarrow Fe(OH)_3 + 5OH^- \quad (2)$$

It has been hypothesized that the ferrate anion destabilizes colloidal particles by the formation of multicharged cationic iron species when Fe^{6+} is reduced to Fe^{3+} (Waite, 1981). It is possible that the reduction of Fe^{6+} to Fe^{3+} results in the formation of intermediate Fe^{5+} and Fe^{4+} hydrolytic species, which may have a greater net, positive charge than the hydrolytic species generated by aluminum, ferric, or ferrous salts (Waite and Gray, 1984). The formation of a greater variety of hydrolysis species would infer that ferrate is more efficient than other inorganic coagulants in reducing the zeta potential of colloidal particles. The various hydrolysis products undergo polymerization while being reduced from Fe^{6+} to Fe^{3+} with the eventual precipitation of ferric hydroxide.

Initial work using potassium ferrate to treat wastewater containing transuranic elements was performed at Los Alamos National Laboratory (Deininger, 1991). The work demonstrated that ferrate was able to lower the gross alpha radioactivity to lower levels than an equivalent or greater amount of ferric sulfate or potassium permanganate. The patent also implies that a potassium ferrate treatment process would assist the facility in meeting future Department of Energy (DOE) 5400.5 regulations while generating less radioactive sludge. The purpose of the work in this paper is to further demonstrate the applicability of potassium ferrate to treat radioactive wastewater and assist a DOE facility in meeting DOE 5400.5 regulatory discharge limits.

The use of potassium ferrate to treat wastewater containing americium and plutonium was demonstrated with jar tests. At the DOE site, the aqueous waste treatment plant currently utilizes 200 to 250 mg/L calcium hydroxide [$Ca(OH)_2$] and 75 to 100 mg/L ferric sulfate [15 to 20 mg/L Fe^{3+}, $Fe_2(SO_4)_3 \cdot 9H_2O$] in each clarifier of a two-clarifier treatment process to treat radioactive wastewaters. The primary function of the wastewater treatment plant is to remove ^{241}Am and $^{238,239}Pu$ from the wastewater before the treated wastewater is discharged to the environment. The incoming waste stream had a gross alpha radioactivity of 37,000 pCi/L. Because the wastewater at this facility is a mixture of radionuclides, DOE Order 5400.5 requires the sum of the ratios of the measured concentration of each radionuclide to its corresponding derived concentration guide value in the treated effluent not to exceed 1.0. According to DOE Order 5400.5, the derived concentration guide value is the ingestion of 730 liters of drinking water over a one-year period that would result in 100 mrem of a specific radionuclide being taken into the body. The derived concentration guide values for the radionuclides present in this waste stream are 40 pCi/L for ^{238}Pu, 30 pCi/L for ^{239}Pu, and 30 pCi/L for ^{241}Am (Radiation Protection of the Public and the Environment, 1990).

Experimental

Reagents and standard solutions

Reagent grade chemicals were used throughout this work. The potassium ferrate(VI) used was 14 to 15 weight-percent (wt%)

TABLE 1
Results of jar tests to find optimum pH for removal of radionuclides using potassium ferrate alpha activity (pCi/L).

Jar test	*Initial alpha activity*	*Results after 1st treatment*	*Results after 2nd treatment*	*Decon. factor*
pH 10.0	3.41×10^6	4.12×10^5	2.93×10^5	12
pH 10.5	3.41×10^6	4.79×10^5	3.13×10^5	11
pH 11.0	2.90×10^6	3.32×10^5	1.00×10^5	29
pH 11.5	3.02×10^6	2.53×10^5	9.75×10^3	310
pH 12.0	3.05×10^6	2.53×10^5	8.50×10^3	359
pH 12.5	2.96×10^6	3.66×10^5	4.83×10^4	61

(TRU/Clear® "4", Analytical Development Corporation, Colorado Springs, CO).

Instrumentation

The jar tests were performed in 2-L "gator" jars (square beakers with siphon hole) or 1,500-mL griffin beakers using a four-vessel jar-test stirrer (Phipps and Bird, Richmond, VA) with continuous speed variation (ASTM, 1990). The concentration of ferrate (FeO_4^{2-}) was monitored using visible spectrophotometry (Kaufman and Schreyer, 1956) using a preprogrammed spectrophotometer (DR/2000, Hach Company, Loveland, CO). A pH-meter (Hach Company) with a combination pH electrode (Orion Research, Boston, MA) was used for all pH measurements.

Gross alpha radioactivity and specific radionuclide analyses were performed on site by site personnel. Gross alpha radioactive analysis was performed on the high-level radioactive samples using a proportional counter (Model PC-5, Nuclear Measurement Corp.). Gross alpha radioactive analysis was performed on the low-level radioactive samples using an internal proportional counter (Canberra Instruments, Meriden, CT). Both procedures follow Standard Methods 7110A (1989) and EPA Method 900.0 (1980) for basic analytical guidelines.

Results and Discussion

Optimum treatment pH determination

The jar tests were single-run, screening tests of tap water spiked with ^{239}Pu and ^{241}Am to provide basic data needed to determine the optimum pH range for utilizing potassium ferrate to remove americium and plutonium from wastewater. The optimum treatment pH range would be determined based on the decontamination factor (the ratio of the contaminant in the untreated waste stream to that in the treated waste stream) obtained from the jar tests. Decontamination factors (DFs) are used to determine the effectiveness of a treatment process instead of percent removal because of the small concentration of radioactive elements usually present in wastewater. Each sample was treated twice with potassium ferrate to determine the optimum treatment pH. In the first treatment step, each 1-L jar test was treated with 5 mg Fe^{6+}. The sample was mixed for one minute at a *G* value of approximately 300 sec^{-1} and allowed to flocculate (*G* value of 20 to 30 sec^{-1}) for 30 minutes. The resulting floc was allowed to settle for 2 hours to simulate actual plant operating conditions. An aliquot of each sample was removed for analysis. The liquid was decanted and treated a second time with 5 mg Fe^{6+} using the same jar test conditions described above. The results from the jar tests are summarized in Table 1. The results indicated that the optimum pH range for removing ^{241}Am and ^{239}Pu from this waste stream with potassium ferrate is 11.5 to 12.0 based on the treatment DF. The results demonstrate using potassium ferrate in a two-stage treatment process will lower the gross alpha activity from 3.0×10^6 pCi/L to less than 10^4.

Wastewater treatment—single-step treatment

The first set of jar tests provided data on whether a single- or two-step treatment procedure would be needed to meet DOE Order 5400.5 discharge regulations for radionuclides. In the first treatment step, each 2-L jar test was treated with 5 mg/L Fe^{6+}. Prior to the addition of K_2FeO_4, the pH of the plant influent was adjusted from 7.9 to a pH range of 11.0 to 12.0 with 50 wt% NaOH. K_2FeO_4 treated samples were mixed for 60 minutes at a *G* value of 20 to 30 sec^{-1}. The resulting floc settled for two hours to simulate actual plant operating conditions.

The results from jar tests 1, 2, and 3 are summarized in Table 2. These results show that 5 mg/L Fe^{6+} used to treat the plant influent in

TABLE 2
Jar test results from one-step treatment of plant influent with 5 mg/L Fe^{6+}.

		Alpha activity after treatment[b]	
Jar test	*Treatment[a] pH*	*Total (pCi/L)*	*Dissolved (pCi/L)*
1	11.0	7000 ± 500	420 ± 130
2	11.5	1900 ± 300	240 ± 100
3	12.0	1500 ± 200	90 ± 70

[a]Initial sample pH was 7.9
[b]Initial sample alpha activity was 37,000 pCi/L ± 2000 pCi/L.

the pH range of 11.5 to 12.0 (jar tests 2 and 3) will lower the gross alpha radioactivity from 37,000 pCi/L to a range of 1,500 to 1,900 pCi/L (decontamination factors of 25 and 19, respectively). The gross alpha radioactivity was lowered to 7,000 pCi/L (DF −5) when the plant influent was treated with 5 mg/L Fe^{6+} at a pH of 11.0 (jar test 1). Filtering the pH 12.0 treated plant influent through a 0.45 μm nylon filter further lowered the gross alpha radioactivity to 90 pCi/L (DF −411). Filtering the pH 11.0 and 11.5 treated samples lowered the gross alpha radioactivity to 420 (DF −88) and 240 pCi/L (DF −154), respectively. However, in order to meet DOE Order 5400.5 discharge specifications, a second treatment step is necessary.

Wastewater treatment—two-step treatment

A second set of jar tests was performed to verify whether or not a two-step treatment would meet DOE 5400.5 specifications. In the first treatment step, the pH of the three 2-L plant influent samples was raised to a pH of 11.0, 11.5, and 12.0 with 50 wt% NaOH. After pH adjustment, each 2-L jar test was treated with 5 mg/L Fe^{6+}. The sample was mixed for one minute at G value of approximately 300 sec^{-1}. The sample was mixed for one hour at a G value of 20 to 30 sec^{-1}. The resulting floc was allowed to settle for two hours to simulate actual plant operating conditions. The majority of treated liquid was siphoned into another 2-L jar to be treated a second time with 5 mg/L Fe^{6+} using the same treatment conditions just described. If necessary,

TABLE 3
Jar test results from two-step treatment of plant influent with 5 mg/L Fe^{6+}.

		Alpha activity after treatment[b]	
Jar test	*Treatment*[a] *pH*	*Total (pCi/L)*	*Dissolvd (pCi/L)*
4	11.0	90 ± 70	<40
5	11.5	60 ± 60	<40
6	12.0	<40	<40

[a]Initial sample pH was 7.9.
[b]Initial sample activity was 37,000 pCi/L ± 2000 pCi/L.

the pH of the once-treated sample was adjusted to the pH used in the first treatment step.

The results from the jar tests are summarized in Table 3. The results show that utilizing a two-step treatment procedure to treat the DOE site wastewater will lower the gross alpha radioactivity from 37,000 pCi/L to less than 90 pCi/L (FD – 410). The treated samples were filtered with 0.45 μm nylon filter, which lowered the gross alpha radioactivity to less than 40 pCi/L (D.F. –925), the detection limit for gross alpha radioactivity at the DOE site.

Conclusions

Jar tests were performed to evaluate the applicability of using potassium ferrate(VI) [K_2FeO_4] to treat radioactive wastewater at a DOE facility for the removal of americium and plutonium. A two-step K_2FeO_4 treatment utilizing 5 mg/L Fe^{6+} in each treatment step proved to lower the gross alpha radioactivity from 37,000 pCi/L to less than 40 pCi/L at a treatment pH of 11.0 to 12.0. The use of potassium ferrate ensures that the facility would be able to meet the DOE 5400.5 regulations for the discharge of radioactive wastewaters to the environment. With tighter discharge regulations becoming more prevalent, potassium ferrate offers an alternative to current treatment chemicals to meet wastewater discharge treatment goals.

Registry No. potassium ferrate(VI), 13718-66-6xU

References

Amartharajah, A., and Tambo, N. (1991) Mixing in Coagulation and Flocculation, AWWA, 3.
American Society for Testing and Materials (1990) 1990 Annual Book of ASTM Standards. ASTM, Philadelphia, PA.
Deininger, J. P. (1991) U.S. Patent 4,983,306.
DeLuca, S. J., Chao, A. C., and Smallwood, Jr., C. (1983) Removal of Organic Priority Pollutants by Oxidation-coagulation. *J. Environ. Eng.*, **109,** 36.
DeLuca, S. J., Contelli, M., and DeLuca, M. A. (1992) Ferrante vs. Traditional Coagulants in the Treatment of Combined Industrial Wastes. *Water Sci. Technol.*, **26,** 2077.
Haas, C. N. (1990) Water Quality and Treatment. McGraw-Hill, New York, 881.
Kaufman, Z. G., and Schreyer, J. M. (1956) Spectrophotometry of the Ferrate(VI) Ion in Aqueous Solution. *Chemist Analyst,* **45,** 22.
Murmann, R. K., and Robinson, P. R. (1974) Experiments Utilizing FeO_4^{2-} for Purifying Water. *Water Res.* (G.B.), **8,** 543.
Radiation Protection of the Public and the Environment (1990) DOE Order 5400.5. U.S. Government Printing Office, Washington, D.C.
Schreyer, J. M., and Ockerman, L. T. (1951) Stability of the Ferrate(VI) Ion in Aqueous Solutions. *Anal. Chem.*, **23,** 1313.
Standard Methods for the Examination of Water and Wastewater (1989) 17th Ed., American Public Health Association, Washington, D.C.
U.S. Environmental Protection Agency (1980) Prescribed Procedures for Measurement of Radioactivity in Drinking Water. EPA-600/4-80-032, Washington, D.C.
Waite, T. D. (1981) Chemistry in Water Reuse, Ann Arbor Science, Ann Arbor, Mich., 543.
Waite, T. D., and Gray, K. A. (1984) Oxidation and Coagulation of Wastewater Effluent Utilizing Ferrate(VI) Ion. *Stud. Environ. Sci.*, **23,** 407.

Here is a sample critique of the Potts and Churchwell article.

The authors of this article claim to have demonstrated a method for removing enough of the radionuclides plutonium and americium from industrial wastewater to satisfy Department of Energy (DOE) regulations. The method uses potassium ferrate, K_2FeO_4 in an alkaline solution to develop a floc, a light precipitate, to carry the ions of ^{238}Pu, ^{239}Pu and ^{241}Am to the bottom of the containing vessel. These radionuclides are all potent alpha emitters and their removal from wastewater is essential.

The work described is a continuation of the work of other authors in this area. Removal of these radionuclides from wastewater has been attempted using aluminum and iron salts. In addition, some initial work using potassium ferrate has already been performed and a theory of the mechanism of the removal of ions from solution using potassium ferrate has been proposed.

The experiments reported explore the optimum conditions for removal of the radionuclides and go on to determine what conditions are necessary to meet the DOE regulations for wastewater discharge. The principal parameter tested for maximizing the effectiveness of the method is pH. The pH's for six experiments ranged from 10.0 to 12.5. The level of alpha particle emission was measured for water samples to which ^{239}Pu and ^{241}Am had been added. The samples were then treated with potassium ferrate, the liquid was drawn off and the level of alpha emission was measured. The process was then repeated. Calculating ratios of initial and final alpha emission showed that the best separation occurred at a pH of 11.5 to 12.0. Experiments run within this pH range with actual wastewater samples showed that removal of radionuclides using two treatments of potassium ferrate was sufficient to meet DOE standards.

The article presents a method for cleaning industrial wastewater containing alpha emitters, clearly an important goal, and the data presented are convincing. Of particular interest was the determination of a very narrow range of pH for maximum effectiveness. Since only total alpha emission was measured, the paper did not in any way explore whether the method was more effective for one radionuclide than another. It would be useful and interesting to know this. Furthermore, the effect of temperature, usually a key parameter for chemical processes, was not determined.

This critique first presents a summary of the article. It then goes on to give the results with an indication of their importance. The most interesting of the results are specified, and the critique then mentions experiments not performed in this paper, suggesting work for the future.

6

Writing Literature Reviews

A chemistry textbook is intended to present a limited but up-to-date and "correct" description of the subject. By "correct," we mean that the information and theories contained in the textbook are currently accepted by the overwhelming majority of chemists. Of course, new discoveries and more persuasive theories will eventually modify some of what is now considered "correct."

Chemistry textbooks fulfill a useful function by pulling together in one place many of the established facts and accepted theories that define chemistry as a field of study. At the same time, however, precisely because chemistry textbooks describe established facts and accepted theories, they often leave out the processes that led chemists to establish particular facts and to accept particular theories in the first place. That is, they tend to leave out the processes of interpretation by which chemists seek to persuade the larger community of chemists of the relevance of new facts or the significance of new theories.

For this reason, students often think of chemistry as the body of knowledge chemists have discovered and that learning chemistry is largely a matter of acquiring the established facts. This is unfortunate because it leaves out learning about that key step by which chemists establish the facts through persuasive interpretations. To put it another way, learning to think like a chemist requires understanding how chemists make interpretations that are accepted by the community of chemists.

In this chapter, you will learn how chemists establish facts and theories through interpretation, and you will be introduced to the literature review—a form of scientific writing that explains the current state of interpretation concerning a particular topic.

INTERPRETATION IN CHEMISTRY

The history of chemistry is the story not just of notable discoveries but also of contradictory views, controversy, and arguments about interpretation. In fact, the controversies in chemistry are rarely about the results a chemist has obtained; chemistry demands that the details of experiments be described and results and new discoveries can always be checked. The controversies in chemistry are about interpretation.

Scientific Controversy

An interesting example of controversy over interpretation involves the fact that tartaric acid, a crystalline substance found in grapes, can exist in very similar but distinguishable forms. In 1874, Jacobus van't Hoff and Joseph Le Bel independently arrived at an interpretation of this involving a geometrical argument. They assumed that the four bonds from a carbon atom point to the corners of a regular tetrahedron, and if four different groups are bonded to the same carbon atom, there can be two different arrangements that are mirror images of each other. This has been the accepted explanation for decades, but at the time it stirred up great controversy. Consider the animosity shown by the powerful German chemist H. Kolbe (1877) toward van't Hoff.

> A Dr. J. H. van't Hoff, of the veterinary school at Utrecht, finds as it seems, no taste for exact chemical investigation. He has thought it more convenient to mount Pegasus (obviously loaned by the veterinary school) and to proclaim in his "Chemistry in Three-Dimensions" how during his bold flight to the top of the chemical Parnassus, the atoms appeared to him to have grouped themselves throughout universal space. (*J. Prakt. Chem.*, **15,** 473)

Kolbe and many others were clearly used to writing chemical structures only in two dimensions, as on paper. In fact, atoms do group themselves "throughout universal space" and the theory of van't Hoff and Le Bel is now one of the cornerstones of the study of chemical structure.

Priestley and Lavoisier

Differences in interpretation show in other examples. One of the most important examples centers on the discovery of oxygen by Joseph Priestley. Prior to Priestley's discovery, a central theory of chemistry, which was widely accepted, involved the existence of phlogiston. This mythical substance was considered to be the part of matter that was responsible for combustion and could be seen to depart in the form of a flame when something burned. Priestley had been a steadfast believer in the phlogiston theory and saw his discovery of his new gas to be the definitive proof of phlogiston's existence. In his interpretation, oxygen was air from which all of the phlogiston had been removed. The total lack of phlogiston gave this "air" an intense appetite for removing phlogiston from other substances. To Priestley this was why it supported combustion so readily and he called it "dephlogisticated air." In fact, Priestley believed in phlogiston for the rest of his life, outliving essentially all of his fellow believers.

Antoine Lavoisier made quite a different interpretation of Priestley's results. He saw that oxygen was an element and that combustion of a substance represented the combination of oxygen with the substance. This interpretation was one of the principal developments of modern chemistry and started a revolution in chemical thinking.

The following excerpt by Thomas Henry Huxley discusses the differences in Priestley's and Lavoisier's interpretations. Huxley himself was a famous biologist and philosopher and was a staunch supporter of Darwin's theory of evolution. Huxley's grandson Aldous is famous as the author of *Brave New World*, one of the most celebrated novels of life in the future.

Joseph Priestley

Thomas Henry Huxley

That Priestley's contributions to the knowledge of chemical fact were of the greatest importance, and that they richly deserve all the praise that has been awarded to them, is unquestionable; but it must, at the same time, be admitted that he had no comprehension of the deeper significance of his work; and so far from contributing anything to the theory of the facts which he discovered, or assisting in their rational explanation, his influence to the end of his life was warmly exerted in favour of error. From the first to last, he was a stiff adherent of the phlogiston doctrine which was prevalent when his studies commenced; and, by a curious irony of fate, the man who by the discovery of what he called "dephlogisticated air" (oxygen gas) furnished the essential datum for the true theory of combustion, of respiration, and of the composition of water, to the end of his days fought against the inevitable corollaries from his own labours. His last scientific work, published in 1800, bears the title, "The Doctrine of Phlogiston established, and that of the Composition of Water refuted."

When Priestley commenced his studies, the current belief was, that atmospheric air, freed from accidental impurities, is a simple elementary substance, indestructible and unalterable, as water was supposed to be. When a combustible burned, or when an animal breathed in air, it was supposed that a substance, "phlogiston," the matter of heat and light, passed from the burning or breathing body into it, and destroyed its powers of supporting life and combustion. Thus, air contained in a vessel in which a lighted candle has gone out, or a living animal had breathed until it could breathe no longer, was called "phlogisticated."

In the course of his researches, Priestley found that the quantity of common air which can thus become "phlogisticated," amounts to about one-fifth the volume of the whole quantity submitted to experiment. Hence it appeared that common air consists, to the extent of four-fifths of its volume, of air which is already "phlogisticated"; while the other fifth is free from phlogiston, or "dephlogisticated." On the other hand, Priestley found that air "phlogisticated" by combustion or respiration could be "dephlogisticated," or have the properties of pure common air restored to it, by the action of green plants in sunshine. The question, therefore, would naturally arise—as common air can be wholly phlogisticated by combustion, and converted into a substance which will no longer support combustion, is it possible to get air that shall be less phlogisticated than common air, and consequently support combustion better than common air does?

Now, Priestly says that, in 1774, the possibility of obtaining air less phlogisticated than common air had not occurred to him. But in pursuing his experiments on the evolution of air from various bodies by means of heat, it happened that on the 1st of August, 1774, he threw the heat of the sun, by means of a large burning glass (magnifying glass) which he had recently obtained, upon a substance which was then called *mercurius calcinatus per se*, and which is commonly known as red precipitate (mercury II oxide, HgO).

> I presently found that, by means of this lens, air was expelled from it very readily. But what surprised me more than I can well express, was that a candle burned in this air with a remarkably vigorous flame.

Priestley obtained the same sort of air from red lead, but, as he says himself, he remained in ignorance of the properties of this new kind of air for seven months, or until March 1775, when he found that the new air behaved with "nitrous gas" (nitrogen monoxide [$2NO (g) + O_2 (g) \rightarrow 2NO_2 (g)$]) in the same way as the dephlogisticated part of common air does; but that, instead of being diminished to four-fifths, it almost completely vanished, and therefore, showed itself to be "between five and six times as good as the best common air I have ever met with." As this new air thus appeared to be completely free from phlogiston, Priestley called it "dephlogisticated air."

What was the nature of this air? Priestley found that the same kind of air was to be obtained by moistening with the spirit of nitre (nitric acid) any kind of earth that is free from phlogiston (for example, CaO), and applying heat ($CaO + 2HNO_3 \rightarrow H_2O + Ca(NO_3)_2$); $Ca(NO_3)_2 \rightarrow O_2 + Ca(NO_2)_2$; and consequently he says: "There remained no doubt on my mind but that the atmospherical air, or the thing that we breathe, consists of the nitric acid and earth, with so much phlogiston as is necessary to its elasticity, and likewise so much more as is required to bring it from its state of perfect purity to the mean condition in which we find it."

It would have been hard for the most ingenious person to have wandered farther from the truth than Priestley does in this hypothesis; and, though Lavoisier undoubtedly treated Priestley very ill, and pretended to have discovered dephlogisticated air, or oxygen, as he called it, independently, we can almost forgive him when we reflect how different were the ideas which the great French chemist attached to the substance which Priestley discovered.

They are like two navigators of whom the first sees a new country but takes clouds for mountains and mirage for lowlands; while the second determines its length and breadth, and lays down on a chart its exact place, so that, thenceforth, it serves as a guide to his successors, and becomes a secure outpost whence new explorations may be pushed.

Nevertheless, as Priestley himself somewhere remarks, the first object of physical science is to ascertain facts, and the service which he rendered to chemistry by the definite establishment of a large number of new and fundamentally important facts is such as to entitle him to a very high place among the fathers of chemical science.

Writing Assignments

1. Write a summary of Priestley's experimental results as they are presented in this passage and explain how Priestley interpreted these results to uphold the doctrine of phlogiston.

2. Write a short essay that offers a possible explanation of how such an accomplished scientist as Priestley could have missed the significance of his own discovery. What does this tell you about the way that science is performed?
3. What conclusions does Huxley draw from this episode in the history of chemistry? Write a short essay about what Huxley sees as the relative contributions of Priestley and Lavoisier to the discovery of oxygen.

Bragg's Interpretation

Another example of an interpretation that was contrary to the intuition of the time involved the use of X-rays to determine the structures of sodium chloride and closely related salts, the work of the father and son team, William Henry Bragg and William Lawrence Bragg. It had previously been generally assumed that a compound such as NaCl would have an identifiable molecule containing one sodium atom and one chlorine atom. The Braggs' structures for these compounds have no such molecules; in NaCl each sodium ion is simply surrounded by six chloride ions and each chloride ion is surrounded by six sodium ions.

In the following excerpt from one of William Lawrence Bragg's papers, he makes the critical analysis interpreting the following experimental observations: KBr and KI have very similar X-ray diffraction patterns; the diffraction pattern of KCl is quite different; and the diffraction pattern of NaCl (rock salt) is intermediate between the patterns of KBr and KI and that of KCl.

The Structure of Some Crystals

W. L. Bragg

On comparing the evidence as to the nature of the diffracting systems in these crystals of sodium chloride, and of potassium chloride, bromide, and iodide, it would seem that a very simple explanation of their curious difference may be arrived at when it is considered that in each case diffraction is caused by two different atoms, and that the relative efficiencies of the two vary from crystal to crystal. Any explanation of these differences would be an extremely improbable one which did not assume a similar structure for the whole group of alkaline halides, for these crystals resemble each other very closely in their properties. Yet it has been seen that the space lattice of diffracting points is the simple cubic one in KCl; it is the face-centered cubic lattice in KBr and KI, and that in the case of NaCl the diffracting point system is in some way intermediate between the two space lattices.

It is reasonable to assume provisionally that the weight of the atom in the main defines its effectiveness as a diffracting centre, and that two atoms of equal weight are equally effective. In the case of potassium chloride the atoms of potassium and chlorine, of atomic weight 39 and 35.5 respectively, are sufficiently close in atomic weight to act as identical diffracting centres. For rock salt this is no longer true; the atomic weight of sodium and chlorine differ considerably (35.5 to 23), and complications are introduced into the simple pattern characteristic of potassium chloride. In potassium bromide and iodide one atom preponderates so greatly over the other in atomic weight that the diffracting system consists practically of atoms of one kind only, and the pattern can again be assigned to a simple space lattice, but one which is of a different nature to that of potassium chloride. Yet the atoms of alkaline metal and halogen have precisely the same arrangement in all these cases. Let us distinguish between two kinds of diffracting points by calling them black and white. Then the points must be arranged in such a way that—

1. There are equal numbers of black and white.
2. The arrangement of points black and white taken all together is that of the one cubic lattice.
3. The arrangement of blacks alone or of whites alone is that of a different cubic lattice.

An arrangement which gives this result is shown in fig. 10.

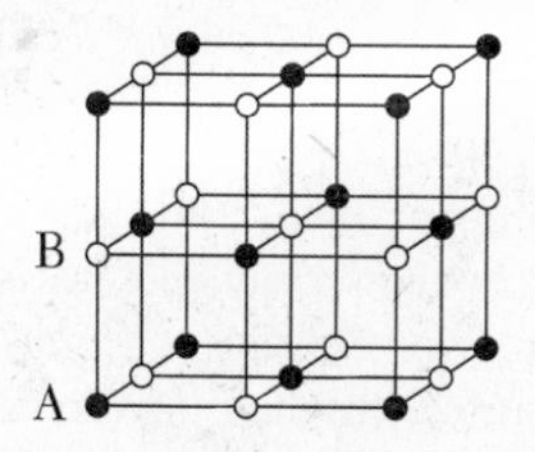

In this diagram we may associate black centres with the alkaline metal, and white with the halogen, or *vice versa.* The space lattice formed by the whites is the same as that formed by the blacks, being in each case the face-centered cube. If black and white centres become identical, as in potassium chloride, the diffracting lattice becomes the simple cubic one.

Writing Assignments

1. Write a summary that describes in your own words the main points of Bragg's logic in this passage.
2. Bragg won the Nobel Prize in 1915 for determining the structure of one of the simplest compounds in existence, NaCl, having only two independent ions to locate. Since that time, thousands of structures have been determined including molecules having tens of thousands of atoms that have been precisely located. Were the Nobel awarders of 1915 too hasty in giving Bragg the Nobel Prize and should they have waited to see how much more complex would be the structures that were determined? Write a short essay that explains your answer.

The Problem of Valence

An example of successive advances in chemical interpretation is provided by the following two articles on the theory of the structure of chemical compounds. The first excerpt is the initial substantive paragraph of a paper by Bray and Branch, in which the authors are struggling with molecular structure and the concept of *valence*. In the second article, published shortly thereafter, Lewis answers a number of the uncertainties in the earlier paper and, in fact, establishes concepts that are used to this day.

The word "valence" is a traditional term in chemistry. Valence has to do with the power or ability of an element to combine with other elements to form compounds. Thus an atom can be assigned a number to indicate its combining power and this number is called the "valence" of the atom. However, as Bray and Branch point out, it is possible for the word to mean more than one thing.

In particular, Bray and Branch deal with two common uses of the word "valence," one by the organic chemists of their time and one by the inorganic chemists. They give these two different meanings the names *polar* number and *total* valence number. Ammonium chloride is given as an example of the uses of the two different meanings.

Before you read this first excerpt, write out the Lewis dot structure for ammonium chloride and be sure that you understand the nature of this compound.

Valence and Tautomerism

W. C. Bray and G. E. K. Branch

Polar Number and Total Valence Number

In the first place it is necessary to differentiate the two ideas embodied in the term valence number, each of which is frequently emphasized to the exclusion of the other. We suggest that these be distinguished by the names *polar* number and *total* valence number. The latter alone, is met with in elementary organic chemistry; and the former is becoming more and more popular in elementary inorganic chemistry.

The difference between the two ideas may be illustrated by means of ammonia and ammonium chloride. Many organic chemists insist that the valence numbers of the nitrogen in these compounds are 3 and 5 respectively, and thus emphasize the total number of valence bonds of the nitrogen, as illustrated by the following structural formulas:

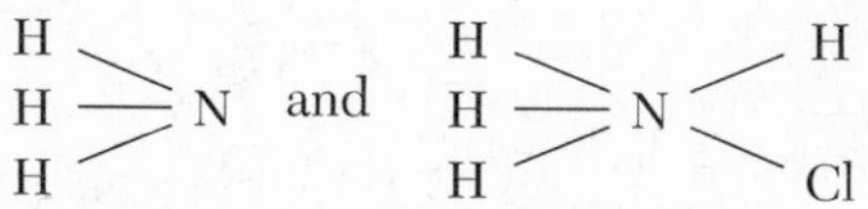

On the other hand, many inorganic chemists consider that it is more important to emphasize the relationship between ammonia and ammonium chloride, and insist that the valence number of the nitrogen is −3 in both cases. From this point of view the valence number of hydrogen in its compounds is in general +1, that of chloride chlorine is −1, and that of nitrogen is −3 in ammonia, ammonium ion, or any ammonium salt.

Many of the misunderstandings which have arisen in this matter seem to us to be due to the failure to recognize that two separate ideas are involved, namely, that of polarity and that of the total number of bonds. The latter idea has probably the better right to the name valence; but the polarity idea is of so great value in elementary inorganic chemistry that it is doubtful if the term valence could be ousted from the textbooks. It therefore seems advisable to retain the term valence for the two ideas, and to use the distinguishing terms *polar* number and *total* valence number whenever necessary. The valence of nitrogen in ammonium chloride can then be completely described as $(-3, 5)$ where, for convenience, the *polar* number is placed first.

Writing Assignments

1. Write a short essay that defines the problem Bray and Branch address. Speculate on why organic and inorganic chemists used different terms to discuss valence. Explain these differences. Is the distinction between *polar* number and *total* valence number useful? Why or why not?
2. How would this paper change if Bray and Branch knew then what you know now about ammonium chloride? Write a short essay that explains your answer.

The following excerpt is from a paper by Gilbert Newton Lewis with the imposing title, "The Atom and the Molecule." In this paper Lewis moved on from the ideas of Bray and Branch and established the basis of much of the modern view of structure and bonding.

The Atom and the Molecule

G. N. Lewis

The Cubical Atom

A number of years ago, to account for the striking fact which has become known as Abegg's law of valence and countervalence, and according to which the total difference between the maximum negative and positive valences or polar numbers of an element is frequently eight, I designed what may be called the theory of the cubical atom.

This theory, while it has become familiar to a number of my colleagues, has never been published, partly because it was in many respects incomplete. Although many of these elements of incompleteness remain, and although the theory lacks today much of the novelty which it originally possessed, it seems to me more probable intrinsically than some of the other theories of atomic structure which have been proposed, and I cannot discuss more fully the nature of the differences between polar and nonpolar compounds without a brief discussion of this theory.

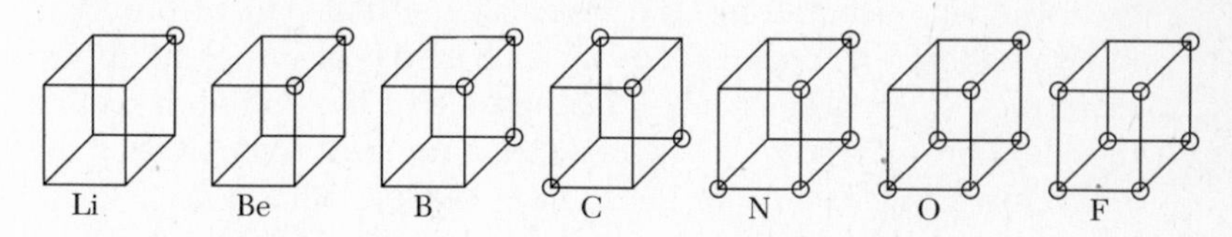

The pictures of atomic structure which are reproduced in Fig 2, and in which the circles represent the electrons in the outer shell of the neutral atom, were designed to explain a number of important laws of chemical behavior with the aid of the following postulates:

1. In every atom is an essential *kernel* which remains unaltered in all normal chemical changes and which possesses an excess of positive charges corresponding in number to the ordinal number of the group in the periodic table to which the element belongs.
2. The atom is composed of the kernel and an *outer atom* or *shell*, which, in the case of the neutral atom, contains negative electrons equal in number to the excess of positive charges of the kernel, but the number of electrons in the shell may vary during chemical change between 0 and 8.
3. The atom tends to hold an even number of electrons in the shell, and especially to hold eight electrons which are normally arranged symmetrically at the eight corners of a cube.
4. Two atomic shells are mutually interpenetrable.
5. Electrons may ordinarily pass with readiness from one position in the outer shell to another. Nevertheless they are held in position by more or less rigid constraints, and these positions and the magnitude of the constraints are determined by

the nature of the atom and of such other atoms as are combined with it.

6. Electric forces between particles which are very close together do not obey the simple law of inverse squares which holds at greater distances.

Some further discussion of these postulates is necessary in order to make their meaning clear. The first postulate deals with the two parts of the atom which correspond roughly with the inner and outer rings of the Thomson atom. The kernel being that part of the atom which is unaltered by ordinary chemical change is of sufficient importance to merit a separate symbol. I propose that the common symbol of the element printed in a different type be used to represent the kernel. Thus **Li** will stand for the lithium kernel. It has a single positive charge and is equivalent to pure lithium ion Li^+. **Be** has two positive charges, **B** three, **C** four, **N** five, **O** six and **F** seven.

We might expect the next element in the series, neon, to have an atomic kernel with eight positive charges and an outer shell consisting of eight electrons. In a certain sense this is doubtless the case. However, as has been stated in Postulate 3, a group of eight electrons in the shell is extremely stable, and this stability is the greater the smaller the difference in charge between the nucleus and this group of eight electrons. Thus in fluoride ion the kernel has a charge of +7 and the negative charge of the group of eight electrons only exceeds it by one unit. In fact in compounds of fluorine with all other elements, fluorine is assigned the polar number −1. In the case of oxygen, where the group of eight electrons has a charge exceeding that of the kernel by two units, the polar number is considered to be −2 in nearly every compound. Nitrogen is commonly assumed to have the polar number −3 in such compounds as ammonia and the nitrides. It may be convenient to assign occasionally to carbon the polar number −4, but it has never been found necessary to give boron a polar number −5, or beryllium −6, or lithium −7. But neon, with an inner positive charge of 8 and an outer group of eight electrons, is so extremely stable that it may, as a whole, beregarded as the kernel of neon and we may write **Ne** =Ne.

The next element, sodium, begins with a new outer shell and **Na** = Na^+, **Mg** = Mg^{++}, and so on. In my original theory I considered the elements in the periodic table thus built up, as if block by block, forming concentric cubes. Thus potassium would be like sodium except that it would have one more cube in the kernel. This idea, as we shall see, will have to be modified, but nevertheless it gives a concrete picture to illustrate the theory.

Writing Assignments

1. Write a short essay that compares Lewis's paper with the paper published three years earlier by Bray and Branch. Lewis's paper is considered a landmark work while that of Bray and Branch is closer to a historical curiosity. Is that a fair judgment? Why or why not?
2. Write a short essay that uses our current terminology, including such terms as valence electrons and noble gas electron configuration, to explain what Lewis means by phrases we don't normally use such as "kernel" and "outer rings." Compare our symbols and the kinds of symbols that he uses. Is one set of terminology and symbols better than the other? Explain your answer.
3. Write a short essay that presents your view of Lewis's notion of building up atoms consisting of concentric cubes. Does this notion strengthen or weaken the paper as far as you are concerned? Discuss Lewis's notion in terms of your own understanding of the nature of the atom.

WRITING THE LITERATURE REVIEW

In many respects, a chemistry textbook is a literature review without reference to the chemists responsible for the research and interpretation that define a particular topic in chemistry. Accordingly, a literature review is a good way for students of chemistry to go beyond their textbooks to reconstruct the interpretive work chemists have done to establish facts and theories concerning a particular topic or area in chemistry. Literature reviews do not simply report the established facts, as textbooks do. Rather, they explain how the facts have been established and what issues of controversy remain within a particular field of investigation. Moreover, literature reviews, unlike textbooks, make judgments about the significance of research and interpretation. As you have seen in the earlier sections of this chapter, ideas are challenged and concepts change in chemistry. A literature review is a way of accounting for these changes.

Here are the steps to take in writing a literature review.

Getting Started: Choosing a Topic

Pick a topic that draws on relatively recent research and about which there are still some disagreements and unresolved issues. The following topics offer areas of study where the facts are reasonably established but where there may still be room for alternative interpretations. You can choose one of these or propose similar topics. You may want to consult with your instructor about your choice of topic.

- The use of chemical themes on stamps. (A leading reference is *Chem. Eng. News* **1990**, Feb 12, p. 21.)
- The chemistry on Titan, one of the moons of Saturn. (Start your literature review by looking up the work of C. Sagan and R. Hanel.)
- The chemistry of spider silk.
- Chemical aspects of the Shroud of Turin.
- Chemical studies of the Vinland Map.
- The passage of mercury through the food chains in the ocean and in fresh water.
- The evidence that chlorofluorocarbons, CFC's, are depleting the ozone layer.
- The nuclear chemistry involved in the Chernobyl disaster.
- Acid rain.
- Natural and synthetic rubber.
- The crystals of common minerals.

We will offer suggestions about how to find sources on your topic on pages 134–142.

Understanding Your Audience and Purpose

While it is true that your instructor is sure to read your literature review, he or she may not be the best audience to keep in mind as you're reading and writing, in part because your instructor will be grading your literature review and in part because your instructor may seem to you to know about your topic already. If you consider that the purpose of a literature review is to explain the current state of ongoing work in a particular area of chemistry, then it makes sense to think of your audience as someone interested but not necessarily as knowledgeable as you are about that area. Classmates who are serious about chemistry are a good audience to write for. If you write for them you'll be writing to colleagues who can learn from what you are writing.

Exploring Your Topic

As you begin to read through the literature on your topic, you should keep two main questions in mind: What is the central problem the researchers are investigating? How do the researchers differ in their approach to an understanding of this central problem?

Looking for the central problem in your literature review should lead you to ask why the researchers are raising the particular questions they are trying to answer. As you do your reading of the literature, stop periodically to write and then update a summary of the main problem and the central issues.

At the same time, as you read, try to identify if significantly different positions or interpretations are being presented by different scientists working in the area. It is important to be aware of the level of controversy that exists in the research. If differences emerge in your reading, keep notes on the main points of controversy.

Planning Your Literature Review

Once you have completed your reading, review your notes and begin to plan your literature review. Most literature reviews are organized into four sections:

- Introduction,
- Review,
- Conclusion, and
- References.

Understanding the function of each of these sections will help you to organize your material and to outline your literature review.

We will now consider the first three sections as they would appear in the completed literature review.

The topic of the sample literature review we'll be using is the incorporation of lead from gasoline into the human body. Searching the literature will uncover numerous references to the subject of environmental lead and its effects or presumed effects on humans. However, the study that appears to be the most definitive on gasoline lead and its absorption by humans is the isotopic lead experiment (ILE) which was conducted in the late 1970s in the area of Turin, Italy.

Introduction

The purpose of the Introduction is two-fold. First it should define the problem the researchers are addressing and indicate its significance. Second, the Introduction should then forecast for the reader what research will be reviewed.

Notice the differences in the way these two opening paragraphs define the central problem of the literature review.

EXAMPLE ONE

The isotopic lead experiment (ILE) was performed in the region of Turin, Italy. The ratio of ^{206}Pb to ^{207}Pb was determined in the blood of residents of the city and the surrounding country.

EXAMPLE TWO

Lead has been introduced into gasoline for years to improve the burning characteristics or octane number. Lead is known to have damaging effects on humans, and it can be expected that at least some of the lead in gasoline will be introduced into the atmosphere. Therefore, it is imperative that the connection between lead in gasoline and in human blood be determined. Such studies have been performed since the increased environmental consciousness of the early 1960s.

Notice that the opening sentences in Example One fail to define for the reader the central problem that the literature review is considering. It doesn't really develop the context of the experiments to be described or show the overall significance of the work to be performed. Furthermore, this beginning places too much importance on where the work was done, and it describes the results of research before it has adequately explained why the research was undertaken.

On the other hand, the few sentences in Example Two do define the problem clearly, indicate its significance, and explain why researchers have addressed the problem. Moreover, this version gives an idea of what the literature review is going to be about without giving details that are not important at this time, such as the experimental results.

The second step in the Introduction is to indicate the nature of information to be included in the review. Notice in the following paragraph how the writer explains quite clearly what the literature review is going to do.

This review will survey the important work that has been performed on the correlation between lead in gasoline and lead assimilated into human blood. In particular, it will concentrate on a large-scale experiment, the isotopic lead experiment (ILE).

This paragraph establishes the particular focus of the literature review.

Review

The Review section surveys the work of the researchers, their research methods, and the interpretations they have offered. This section should be more than a simple summary of what you have read. You will want to point out differences among the studies you have reviewed in order to explain how researchers have investigated the central problem and what significance their results hold. Here is a well-written opening to the Review section.

There have been two fundamental groups of studies on the absorption of environmental lead into the human population. The first group of studies relates lead concentrations in the environment to lead concentrations in human blood.[1,2] The second group of studies involves the use of labeling isotopes.[3-6] In this review we will consider only the isotopic studies that have the potential for identifying the actual sources of the lead incorporated in the blood. Of the isotopic studies, only one study, the isotopic lead experiment, actually labeled the lead in gasoline to allow dependable correlation between gasoline lead and blood lead.

Notice in this paragraph how the writer first distinguishes between the research design in two groups of studies—the first that relates lead concentrations in the environment to lead concentrations in human blood and the second that uses isotopic labeling to identify the sources of lead in human blood. Then the writer draws a distinction among the isotopic studies, asserting that only the isotopic lead experiment enables researchers to correlate gasoline lead and blood lead. In other words, the writer begins the review section with an interpretation that provides a framework to understand and evaluate the work that will be reviewed in greater detail.

The main section of the review goes on to describe the studies that were performed and the results that were obtained.

Lead contains a number of stable isotopes and is an unusual element in that its ratio of these isotopes varies considerably from location to location. For example, the ratio of ^{206}Pb to ^{207}Pb varies from 1.04 for lead from Broken Hill, Australia, to about 1.34 for lead from the Mississippi Valley.[3] Since the ratio of isotopes can easily be determined by mass spectrometry, the location from which a lead sample was gathered can be estimated. Furthermore, determining what fraction of lead in blood came from automobile exhaust could be achieved if all of the lead in the gasoline of a region came from a distant source with an isotopic ratio different than that in the region.

Notice that the writer is setting up the description by explaining the topic from the beginning and not assuming that the reader already knows a lot about it. Thus the description of the element lead proceeds to determining the source of lead in blood. The review then goes on to provide more details,

Lead from Broken Hill was converted to tetramethyl lead by standard industrial methods and this was the only lead introduced into motor fuel in Turin, Italy, and within a radius of 10 km of the city during the course of the experiment. The experimenters stated that they were aware that some error would result from the exhaust of vehicles that had taken on gasoline elsewhere. Samples of gasoline from local retail sources were tested at random to be sure that the gasoline actually contained the isotopic ratio expected.

The writer is explaining one aspect of the experiment. Another aspect is,

Blood samples were drawn twice a week for a period of six weeks from 125 volunteers living within Turin and 97 volunteers living in lightly populated rural areas within 8 km of the city. The urban volunteers included 77 males and 48 females. The rural volunteers included 60 males and 37 females. The ages of the subjects ranged from 19 to 72.

The writer is giving details that will help the reader of the review ascertain the quality of the study.

The mass spectrometer (Consolidated Electrodynamics type 703) used in these experiments was a single focusing instrument with a 30.5 cm radius of curvature, 68° deflection and a 60° sector field.

These experimental details again give the reader who is experienced in the field the opportunity to evaluate the work. Further along, the results are discussed.

```
The principal results of this study are the
percentages of the blood lead that have the iso-
topic distribution of Broken Hill and thus arise
from automobile exhaust. Data for over 2500 deter-
minations were averaged giving standard deviations
of 2% for the urban samples and 1% for the rural
samples.
```

These results will be given in detail before proceeding to the conclusion.

Conclusion

The conclusion section sums up the work reviewed, offers the writer's evaluation of the results, and indicates what further research seems to be called for. It can and should include the writer's interpretation of what work is the most valuable and important. It is not necessary or desirable for the writer to comment on whether he or she happens to find the subject interesting or enjoyable.

Compare the following two Conclusions.

```
EXAMPLE ONE

The problem of lead from automobiles in human
blood is very important since there are so many
cars on the road and we want to improve human
health. The work described in this review shows
that lead from gasoline does get into humans.
Therefore, we should have fewer cars or use gaso-
line with no lead in it.

EXAMPLE TWO

Of all of the studies on the absorption of
leaded gasoline into human blood, the isotopic lead
experiment (ILE) is the most definitive and power-
```

ful. By isolating an area and distributing only gasoline with an isotopic ratio of ^{206}Pb to ^{207}Pb, which is different than other local lead sources, the ILE was able to identify how much gasoline lead actually found its way into human blood. The results were surprising in that the contribution of gasoline lead to blood lead was lower than many people would have expected. Furthermore, the effect was quite different for urban and rural areas. About 24% of the lead in the blood of inhabitants of the large and congested city of Turin came from gasoline. In the rural areas, the percentage of lead from gasoline dropped to about 12%. Clearly there are other important sources of blood lead, especially for rural dwellers.

Notice that Example One makes statements that are too obvious (of course, "we want to improve human health") and draws conclusions that are beyond the consideration of the chemical studies that make up the review ("we should have fewer cars or gasoline with no lead in it"). In other words, this conclusion focuses exclusively on the opinions of the writer instead of on the work being reviewed.

In Example Two, on the other hand, the writer does keep the focus on the studies reviewed. The writer's opinion is included, but it is limited to the writer's considered opinion of the value and significance of the various studies. Moreover, the writer indicates what is "surprising" about the ILE study and suggests the need for further research.

References

All the information in a literature review should be referenced in order to give credit to the original investigators, to demonstrate that the review is complete, and to provide others with key references that they can use in their own research. The referencing of articles varies from journal to journal, but there is an essential, recognizable pattern.

The title of the journal is very often abbreviated for purposes of saving space. The individual journal issues vary in frequency of appearance from every week to four or fewer times a year. Usually one year of issues makes up a *volume*, although some journals may have one volume every six months. The volumes are either numbered consecutively from the first year of publication or have the same number as the year in which they were published. Pages are usually numbered consecutively throughout an entire volume and only the initial issue of a particular volume has a page 1. However, there are a few journals that start every issue with page 1. In these cases the issue number must be given in the reference.

Therefore, the minimum amount of information needed to locate a particular journal article is

- Journal title,
- Volume number,
- Page number, and
- Issue number if the journal starts every issue on page 1.

Here is a sample reference format:

Rosseisky, D. R. *J. Chem. Educ.* **1976,** 53, 617.

In this case, the journal title is an abbreviated version of *Journal of Chemical Education,* the year of publication is 1976, the volume number is 53, and the page number of the article is 617.

A reference is cited by a note in the text. This can be a superscripted number, a number in parentheses, or the author's name and the year of publication. Examples are:

Neutron diffraction was used to verify the structure of copper hydride.[4]

Neutron diffraction was used to verify the structure of copper hydride (4).

Neutron diffraction was used to verify the structure of copper hydride (Goedkoop and Andresen, 1955).

For a literature review and most articles, the references should be collected at the end of the text under the heading, "References" or "Literature Cited." If the citations are numbers, each reference should be listed in order of appearance starting with the number 1 either followed by a period as,

1. Sinn, W. *Int. Arch. Occup. Environ. Health* **1980,** *47*, 93.

or superscripted,

[1]Sinn, W. *Int. Arch. Occup. Environ. Health* **1980,** *47*, 93.

If the same reference is cited later in the paper, use its original number, not a new number.

If the references are cited with the author's name and the year, the references must be listed at the end of the paper in alphabetical order. A paper by Goedkoop and Andresen published in 1955 would be cited as (Goedkoop and Andresen, 1955). If there are more than two authors, give only the name of the first author followed by *et al.* For example, a paper by Chartsias, Colombo, Hatzichristidis and Leydendecker published in 1986 would be cited as (Chartsias *et al.*, 1986)

The reference format of a journal article that is approved by the American Chemical Society looks like this:

4. Goedkoop, J. A.; Andresen, A. F. *Acta Cryst.* **1955**, *8*, 118–119.

Note that the authors' names are written last name first and separated by a semicolon. The name of the journal appears in italics (underlined if done by typewriter) and is not separated from the authors' names by any punctuation. Then comes the year of publication in bold type, again not separated by punctuation. After a comma the volume number appears in italics and lastly, after another comma, comes the page number or numbers in plain type.

If the journal starts every issue with page 1, the issue number must be included and the reference format is:

James, R. D., *Chem. Eng. News* **1994,** 72 (11), 25.

In this reference the year is 1994, the volume is 72, the page number is 25 and the issue number shown in parentheses is 11.

References for books are slightly different.

Mingos, D. M. P.; Wales, D. J. *Introduction to Cluster Chemistry;* Prentice Hall: Englewood Cliffs, NJ, 1990; pp. 117–153.

Note that after the book title, the publisher, the publisher's location, the year of publication, and then the page numbers that have been cited appear. If the entire book is being cited, the page numbers are omitted and a period follows the year. Words such as "Publishers," "Inc.," "Press," and "Company" are omitted from publishers' names.

Often books have different chapters by different authors that were collected by a single editor. In these cases more information must be included in the reference.

Lipscomb, W. N. In *Boron Hydride Chemistry;* Muetterties, E. L., Ed.; Academic Press, New York, 1975; pp. 39–78.

In the reference above, Lipscomb is the author of the chapter of interest and Muetterties is the editor of the book. Note that in book references only the title is in italics and everything else is in plain type.

You may need to cite the thesis of a university student. Give the writer's name, the level of the thesis (M.A., M.S., Ph.D., etc.), the university, and the date that the thesis was presented.

Browne, D. A. Ph.D. Thesis, University of Oklahoma, May 1993.

If you cite a patent, give the investigator's name, the patent number, and the date.

McGimpsey, L. L. U.S. Patent 5 333 987, 1991.

There is also a huge range of U.S. government publications that you may cite, usually available from the Government Printing Office. In addition, there are reports from many foreign governments and agencies. These may come from individual authors or offices or there may be no author listed. There is no regular format for these references and the rule is simply to include as much information in the reference as you can (see reference 6 below).

The conventions for writing references have changed with time and can be different with different journals. For further details on approved conventions for references in American Chemical Society journals consult,

Dodd, J. S. *The ACS Style Guide;* American Chemical Society: Washington, D.C., 1985.

Reference formats for other journals are given in *Chemical Abstracts*, the index for essentially all articles published on chemistry in the world.

The references for the literature review on environmental lead would look like this:

REFERENCES

1. Sinn, W. *Int. Arch. Occup. Environ. Health* **1980,** *47*, 93.
2. Chartsias, B.; Colombo, A.; Hatzichristidis, D.; Leydendecker, W. *Sci. Total Environ.* **1986,** *55*, 275.
3. Manton, W. I. *Arch. Environ. Health* **1977,** *32*, 149.
4. Manton, W. I. *Br. J. Ind. Med.* **1985,** *42*, 168.
5. Facchetti, S. *Mass Spectrom. Rev.* **1988,** *7*, 503.
6. Facchetti, S.; Geiss, F. "Isotopic Lead Status Report," Luxembourg: Commission of the European Communities; 1982, Publication EUR 8352 EN.

Drafting

Once you have written an outline of the first three sections of your literature review, write a first draft. Don't agonize over particular sentences, paragraphs, or sections at this point. What you want to do is get your material written down and in front of you so that you can fine-tune it in a revision. If you get stuck, go to the next point on your outline. Be realistic about the time required to write your first draft. For example, you may want to write the Introduction one day, the Review section the next, and the Conclusion on the third day. Set goals for yourself, but be flexible if you don't meet them. The most important point is to make sure you allow yourself enough time. Starting the night before the literature review is due is way too late.

Revision

Revision is perhaps the most important step in writing your literature review. Make sure you give yourself adequate time to return to your first draft so that you can make the necessary changes. Keep in mind that revision is likely to require you to re-read the studies you're writing about and may even require further research if you find there is an important gap in your review.

Revision means, quite literally, re-seeing what you have already written. In this sense, learning how to read your own writing with a critical eye is crucial to revision. Just as useful is asking a classmate to read what you've written, and to provide feedback on your draft—to indicate where it is clear, where it is not so clear, and what parts need more work. Here are some questions to ask yourself—or ask a classmate to respond to—concerning your draft.

- Does the Introduction define a central problem clearly and indicate what the draft is going to review?
- Does the Review section present the various studies under review in such a way that the reader will be able to identify the central problem and issues of research design and interpretation? Are differences among the studies clearly explained?
- Does the Conclusion sum up the work reviewed in an effective way? Does it offer the writer's interpretation of the significance of the experimental results reviewed? Does it indicate what further problems or research emerge?
- Are there particular sentences or passages that are hard to follow or seem to be out of order? Are there passages in the draft that seem to be underdeveloped, that require more explanation?

Use the responses to these questions to revise your first draft.

Editing and Proofreading

The final step in preparing your literature review is to read it carefully to make sure there are no misspellings or grammatical errors. You may want to rewrite some of the sentences at this point to make them clearer and easier to read. And make sure you check the references so that they are accurate and in proper form.

SEARCHING THE LITERATURE OF CHEMISTRY

Searching the literature of chemistry to find sources for your literature review involves tracing the development of work in a particular area of study. The way to proceed is to start in the present with the most recent publications and work backwards. One of the conventions of scientific literature is that researchers locate their work in relation to previous work, to show how the research they are reporting grows out of and extends or modifies the work that has already been published. For this reason, scientific papers usually include numerous references. Doing a literature search from current publications backwards in time should enable you to reconstruct the work that has most influenced an area of study—you'll begin to notice that some papers are cited more frequently than others—and to see how the ongoing work of science builds on or critiques earlier work.

Searching the Recent Literature

To find the most recent literature, journal articles will be your principal sources of information. These include full papers that have an abstract or summary and full experimental and discussion sections, as well as notes or communications, which are shorter and were intended for rapid publication. Notes or communications are often expanded and published as full papers later. In addition, many journals also publish review articles and these are actually literature reviews. They are fully referenced and can be extremely useful.

Microfilm and Microfiche

Because of the explosion in scientific knowledge and the spiraling costs of producing journals, a few journals are using a "two package concept." In these cases the printed journal contains only a brief synopsis of the work and a few essential figures and important references. The rest of the information is contained in a microfilm or microfiche package normally only subscribed to by libraries. The individual subscriber can order the supplemental pages of the articles in which he or she is interested to be printed up. Because the demands on their space are so great, a number of libraries are now keeping back issues of journals only in microfilm or microfiche versions.

Microfilm and microfiche are read with devices that magnify the image and project it on a screen. Your librarian can show you how to use these devices. In the future we can expect more changes in the nature of scientific journals including some that will be accessible only by a computer network.

Guidelines for the Literature Search

Before starting a literature search, you should have a few points in mind:

1. Know what reference guides (or indices) to chemical literature are available in order to minimize the amount of looking up that you have to do. The most valuable is likely to be *Chemical Abstracts*. (See the discussion below.)
2. Know specifically what your topic is. You will need key words to search through the available indices of chemical literature.
3. Use any information that you have available including people who might have leads on where to look first. In particular, it is very useful to know the names of some chemists who have published recently in the area you are reviewing. You might ask your instructor for information.
4. Decide if you plan to include theses and patents as well as journal articles.
5. Keep a systematic record of your progress. Every time you consult an index, write down the references you find that might be useful. Otherwise you will find yourself looking over and over for the same things in the same index.
6. Stop searching when you have a number of references. Once you start reading, you are likely to find more references.

You can start searching the literature once you have planned out your search according to the criteria above. Most college and university libraries now have computer data bases, which have replaced the old card catalogs and which now include author, title, and subject data for recent journal articles as well as for books. Go to the library computer terminal first and ask for help from your librarian if you need it. You will quite likely find a few lead references upon which you can base your whole literature search. Otherwise there are more comprehensive indices available.

Chemical Abstracts

The field of chemistry is very fortunate that since 1907 the American Chemical Society has published *Chemical Abstracts*, or *CA*, carefully indexed summaries of about 98% of all of the chemical research articles and other relevant information published throughout the world.

CA currently includes abstracts of about 500,000 articles per year from over 12,000 journals published in more than 140 countries. A staff of 1,200 is necessary to maintain this effort. A computer data base including all of the abstracts, and bibliographic information plus complete texts of well over a million articles extends back to 1967. We will consider computer on-line searching of chemical abstracts after we cover the traditional method.

Using *CA* is essentially a matter of understanding its system of indices. Because *CA* "collects" its indices as time passes, the farther back in time you are looking, the fewer indices you need to check. An issue of *CA* comes out every week and each issue contains three indices, a *key word* index, a *patent* index and an *author* index.

- The *key word* index is an alphabetical list of important words or phrases taken from the title and text of the abstract. Some entries from a page of the key word index are shown in Figure 6.1.
- In the *patent* index, patents are grouped numerically under the countries in which they were granted. The countries themselves are arranged alphabetically.
- The *author* index is arranged alphabetically by the last names of the authors.

The abstracts are numbered and an individual abstract is identified by its number and year. Patents are noted by a letter P prior to the abstract number. You must consult the abstract to obtain the actual journal reference.

Six months of *CA* constitutes a volume, and each volume has its own index. The volume indices are more comprehensive than the issue indices. Instead of the key word index, they contain a general subject index, a chemical substance index, a chemical formula index, and an index of compounds with rings of atoms in their structures. There are also author and patent indices as there are for the individual issues.

Before using the volume indices, it is useful to consult the pertinent *CA* Index Guide. This guide provides the information needed to cross-reference the various volume indices. It contains an alphabetical list of subject terms with synonyms, acronyms, and a variety of names including unsystematic common and trade names. The Index

Chlorinated
biphenyl blubber sediment gas chromatog 25436x
biphenyl chlorodibenzodioxin hepatic Ah receptor 256778c
dibenzodioxin dibenzofuran evapn glass filter 25564n
dibenzodioxin dibenzofurran toxicity fish liver 25470d
dihydroxycyclohexadiene chlorocatechol gas chromatog 25429x
phenol cytoxicity 25531z 25532a
Chlorination
bacteria nitric oxide macrophage 29662r
effect benzylpyridinium intramol photo-cyclization 30812w
linalool hydrogen chloride copper P 31723e
oxidative thipoene peroxide hydrochlor-ic acid 31241w
regioselective oxazolidine anodic oxidn 31383u
Chlorine
based substance antimicrobial iontopho-resis 23291x
peracetate sporicide spore suspension biofilm 27597z
tolerance Mesocyclops 25459g
toxicity respiratory tract 25525a
Chlormequat
linseed capsule srouting 25787n
Chloro
nitro salicylamine ethan olamine urea oncomelania P 25873n
ruthenium cyclotadiene dimeric crystal stucture 31680p
Chloroacetophenone
toxicity 25526b
Chloroaniline
degrdn soil temp 25845e
Chloroarom
prepn isomerization zeolite catalyst P 31105e
Chloro benzene
metabolite urine 25492n
methylene transfer oxirane cation radical 30859s

FIGURE 6.1

Part of the key word index from a recent issue of *Chemical Abstracts.* The abstract number follows each entry.

Guide also gives the rules for alphabetizing and includes practical guidelines for literature searches.

Every five years the volume indices are collected into a five-year index. The latest five-year period ends in 1996. No additional information is included in the first-year index beyond that found in the volume indices. However the time saved in consulting only one five-year index rather than ten volume indices is obvious.

Thus the use of *CA* involves looking back through the indices for the appropriate subjects, authors, formulas, and so on. Getting the names of some possible authors can be very helpful. Then, as possible references of abstracts are located you can look them up and read the abstracts to see if the articles are useful. Some sample abstracts are shown in Figure 6.2. If an abstract is promising, you need to get hold of the original article. This will either be on the shelves of your library or available by interlibrary loan. Recent articles will be useful sources of more references to the work that they report, and you can work backwards from a particular reference.

Computer Searching of Chemical Abstracts

Searching *CA* may now also be performed by computer using *Chemical Abstracts Service ONLINE,* or *CAS ONLINE.* This requires a data terminal or microcomputer attached by a communication device to a database vendor. Your library or chemistry department should have this available. Searches can then be done using key word or general subject names, compound names, chemical formulas, or authors' names. Instructions for doing these searches should be available at your location. Detailed instructions for the use of *CA* and *CAS ONLINE* can be found in H. Schulz, tr. E. Mole, *From CA to CA ONLINE,* VCH, New York, 1988.

Other Indices

If you have located the name of a researcher in your desired field you can look him or her up in *Science Citation Index.* This index lists the papers that have given a particular article as a reference. Note that this is a way of working *forward* from a particular reference and will give you more recent references.

There are a number of other chemical indices that can be useful including *Current Abstracts of Chemistry* and *Index Chemicus.* The

116: **262193z Aqucal: calcium hydroxide ready for use.** Ortegat, B. (Cameuse, 5300 Seilles, Fr.). *Trib. Eau* **1991**, 44(554). 9–12 (Fr.) Following a discussion of the application of quicklime and hydrated lime in water and wastewater treatment, the characteristics of Aquacal, a ready for use lime milk delivered in cisterns, is presented.

116: **262194a Experiences with ozone treatment of drinking water in Yugoslavia.** Stankovic, Ivan (Hidroinzenjering Ltd., Energoprojekt, 11070 Belgrade, Yugoslavia). *Ozone: Sci. Eng.* **1992**, 14(2), 101–21 (Eng). O_3 is currently being used in several drinking water treatment plants in Yugoslavia. The new Belgrade water treatment plant Makis is the largest one with 42 kg/h of installed O_3 generating capacity and has been in operation since 1987. This paper describes the main features of O_3 application in drinking water treatment. The exptl. results of Makis pilot-plant investigations and examples of O_3 application in Yugoslavia are presented.

116: **262195b Formation characteristics of formaldehyde and of formaldehyde formed by boiling during the ozonation of surface water.** Yamada, Harumi; Somiya, Isao (Lab. Control Environ. Micropollut., Kyoto Univ., Otsu, Japan 520). *Ozone: Sci. Eng.* **1992**, 14(2), 153–63 (Eng). Lab. studies were conducted to evaluate the effects of ozonization on the formation of HCHO (**I**) in surface water. **I** in chlorinated tap waters also was analyzed. **I** in regular tap water and in boiled tap water was 0–25.8 and 18.0–73.5 μg/L, resp. In surface water, the concns. of **I** and **I** formed by boiling ranged 0–24.0 and 1.5–88.0 μg/L, resp. After ozonization (5 mg O_3/L applied for 5 min) of the surface water, the range increased from 10.0–110 to 21.0–243 μg/L. The relationship between **I** total org. C (TOC) and the amt. of **I** formed was **I** (μg/L) = 21 + 12 × TOC (mg/L). It is suggested that one type of **I** precursors (having the potential to liberate through thermal decompn.) easily forms **I** by chlorination and ozonization.

116: **262196c Economic considerations in the replacement of resins.** Gottleib, Michael; DeSilvia, Francis (Resin Tech Inc., Cherry Hill, NJ USA). *Ultrapure Water* **1992**, 9(3), 50–6, 58 (Eng). The operating efficiency and chem. operating costs are discussed and procedures are established to detn. the economic point at which ion exchange resins from working 2-bed demineralizers should be replaced with new resins.

116: **262197d Treatment methods differ for removing reactive and unreactive silica.** Harfst, William (Harfst and Assoc. Inc., Crystal Lake, IL USA). *Ultrapure Water* **1992**, 9(3), 59–62 (Eng). Prevention of silica-assocd. problems in process water in industry depends on accurate detns. of reactive and unreactive silica, and subsequent treatment of silica prior to use of the water. Reactive and unreactive silica forms and methods to det. and remove these silica species are presented.

FIGURE 6.2

Some abstracts from *Chemical Abstracts.*

use of this is described in more comprehensive guides to the chemical literature such as R. E. Maizell, *How to Find Chemical Information,* 2nd Ed., John Wiley & Sons, New York, 1987; G. Wiggins, *Chemical Information Sources,* McGraw-Hill, Inc., New York, 1991; and Y. Wolman, *Chemical Information, A Practical Guide to Utilization,* 2nd Ed., John Wiley & Sons, New York, 1988.

Searching for Chemical Data

Searching for a specific fact such as the melting point or method of preparation of a compound is a different kind of problem. A handbook is the first place to look for numerical facts such as melting points, vapor pressures, isotopic ratios in naturally occurring elements, and a host of other data. The most familiar handbook is the *CRC Handbook of Chemistry and Physics,* Ed. D. R. Lide, Chemical Rubber Co. Press, Boca Raton, FL. A new edition of this handbook, very similar to the previous edition, appears every year. CRC Press also publishes a number of other more specialized handbooks. Another useful handbook is *The Chemists' Companion, A Handbook of Practical Data, Techniques and References,* Ed. A. J. Gordon and R. A. Ford, Wiley-Interscience, New York, 1973.

The *Merck Index,* published in a new edition every six years or so by Merck & Co., Rahway, NJ, is a premier source of information, particularly on organic compounds. Information included for each compound can include melting and boiling points, solubility, therapeutic information, leading references, and other general facts. Over 10,000 compounds are described and cross-indexed in this single volume.

For methods of preparation and data on organic compounds the best source is *Beilstein's Handbook of Organic Chemistry,* now published by the Beilstein Institute in Frankfort, Germany. This enormous multivolume work was started in 1883 and has been updated with huge supplements since then. It is written in German but its publishers have produced special Beilstein German-English dictionaries to assist English speakers in reading it. A less daunting source of information on organic compounds is *Heilbron's Dictionary of Organic Compounds,* which is also available on-line as *The Dictionary of Organic Compounds.* The inorganic chemistry equivalent of Beilstein is *Gmelin's Handbook of Inorganic Chemistry.* The language here was also German but there has been a changeover to English, which began in 1957. Thermodynamic data are available in *Thermophysical*

Properties of Matter, published by IFI/Plenum, and *JANAF Thermochemical Tables,* published by the National Institute of Standards and Technology. All of these sources and others are available in a good chemistry library. Tips for their use are given in the comprehensive guides to the chemical literature, mentioned above.

Using the INTERNET

Currently there is much excitement about acquiring chemistry data and other information from a global computer information network, the INTERNET. A huge amount of data is continually being posted on this network by many people. This system obviously holds much promise. However, the posted data is unchecked, and at the moment it is difficult for the user to know the reliability of the data that has been retrieved.

Access to the INTERNET is facilitated by a program called "Gopher," which was developed at the University of Wisconsin. This program has been installed at many computing centers and in many cases can be called up simply by typing the letters **gopher** after logging in. Searching within the INTERNET can be facilitated by a program; one such program is called "Veronica." You should contact the computer center at your college or university for directions on accessing and searching the information network.

Another important source of information is the WORLD WIDE WEB. Gary Wiggins from Indiana University Chemistry Library has compiled numerous INTERNET links to chemical resources, called the "Wiggins Compilation," including chemistry book catalogs, data bases, periodicals, software, teaching resources, and newsgroups. For the Wiggins Compilation, the address is **http://www.rpi.edu:80/dept/chem/cheminfo/chemres.html.**

For further information on using the INTERNET and the WORLD WIDE WEB, consult:

F. S. Varveri. "Information Retrieval in Chemistry: Chemistry-Related Anonymous ftp Sites." *J. Chem. Ed.* **1994,** *71,* 872.

H. S. Rzepa, B. J. Whitaker, M. J. Winter. "Chemical Applications of the World-Wide-Web System." *J. Chem. Soc., Chem. Commun.*, **1994,** 1907.

7

Writing Research Proposals

Preparing research proposals is one of the most important kinds of writing that practicing chemists do. Whether chemists work in academia or industry, they are likely to spend a considerable amount of time writing proposals that seek funding and support for the research they wish to pursue. Because research proposals ask for support to pursue a particular experimental approach to a problem, proposals are basically persuasive documents: They seek to persuade their readers, whether they are government funding agencies such as the National Science Foundation and the National Institutes of Health or supervisors in a chemical company, that the research is worthwhile and deserving of support.

As you have seen in the previous chapter, literature reviews explain the current state of work in a particular area of chemistry. Literature reviews often conclude with suggestions about how further research might address unresolved issues in the field of investigation. In this sense, writing research proposals picks up just where the most current literature reviews end—by designing experimental means to explore what is not yet fully understood. Researchers seek to identify problems that emerge from the previous scientific literature and propose ways to tackle these problems in their laboratories.

For this reason, a natural next step in learning to think like a chemist is to use your understanding of how problems are defined, facts established, and theories formulated in the literature of chemistry to design your own research project. In this chapter, we will first offer an overview of proposal writing—how chemists identify and negotiate with likely funding sources to support their research. Next, we will look at how chemists write research proposals that define problems and propose experimental designs to address the problems.

AN OVERVIEW OF PROPOSAL WRITING

Writing a successful proposal that persuades its readers that the research presented is worthy of support is a complicated process. It involves a considerable amount of background research on potential funding sources and negotiations with program officers of the funding agencies. As you will see in this section, chemists do not simply write up what they would like to do in their laboratories and send out proposals that are then either accepted or rejected. A good deal of communication takes place along the way.

Gathering Information on Funding Sources

One of the first things chemists must learn is where to look for information on funding sources to support their research. There are many public agencies and private foundations that support research in chemistry. Chemists typically will do background investigation on these funding sources to see what the goals of the funding source are, what kind of research the funding source has previously supported, and whether there is a good match between what the chemist wants to do and the priorities of the funding source.

Two useful sources of grant funding are published annually. The *Directory of Research Grants*, published by Oryx Press, contains short descriptions of research programs, along with information on application deadlines, amount of funding, special requirements, and the name of contact persons or program officers. The *Guide to Programs*, published by the National Science Foundation (NSF), provides information and guidelines for preparing grant applications to NSF's many programs.

In some cases, such as NSF's Chemistry Division, researchers are invited to submit proposals in most principal areas of chemical investigation. The following announcement appeared in NSF's *Guide to Programs, Fiscal Year 1995*.

Chemistry

The Chemistry Division supports research activities and research infrastructure development in most of the principal subdisciplines of the chemical sciences. However, support is also available from the Divisions of Atmospheric Sciences (atmospheric chemistry), Molecular and Cellular Biosciences (biochemistry, biophysics), Chemical and Transport Systems (chemical engineering), Earth Sciences (geochemistry), and Materials Research (solid-state and polymer science).

The Chemistry Division supports research activities in emerging areas of national interest that cut across "traditional" subdisciplines. These areas include biological chemistry and biotechnology; the chemistry of advanced materials; environmental chemistry, including research in greenhouse gas dynamics, and the program in Environmentally Benign Chemical Synthesis and Processing jointly supported with the Engineering Directorate (see brochure NSF 92-13 for more information); high performance computing and communications; and advanced manufacturing, including fundamental research underpinning chemical and pharmaceutical manufacture. Many of these activities are part of research programs that are coordinated through the National Science and Technology Council.

Research in subdisciplinary areas is also a vital part of the Chemistry Research Project Support investment portfolio. These areas include the following:

- **Analytical and Surface Chemistry**—Supports fundamental chemical research directed toward the characterization and analysis of all forms of matter. This program includes studies of elemental and molecular macrocomposition, and of microstructure of both bulk and surface domains. Investigations designed to probe the interphase region between different forms of matter are the responsibility of this program.
- **Inorganic, Bioinorganic, and Organometallic Chemistry**—Supports research on the synthesis, structure, and reaction mechanisms of molecules containing metals, metalloids, and nonmetals, encompassing the entire periodic table of the elements. Included are studies of stoichiometric and homogeneous catalytic chemical reactions; bioinorganic and organometallic reagents and reactions; and the synthesis of new inorganic substances with predictable chemical, physical, and biological properties. Such research provides the basis for understanding the function of metal ions in biological systems, for understanding the synthesis of new inorganic materials and new industrial catalysts, and

for systematic understanding of the chemistry of most of the elements in the environment.

- **Organic Chemical Dynamics**—Supports research on the structures and reaction dynamics of carbon-based molecules, metallo-organic systems, and organized molecular assemblies. Research includes studies of reactivity, reaction mechanisms, and reactive intermediates, and characterization and investigation of new organic materials. Such research provides the basis for understanding and modeling biological processes, and for developing new or improved theories relating chemical structures and properties.
- **Organic Synthesis**—Supports research on the synthesis of carbon-based molecules, metallo-organic systems, and organized molecular assemblies. Research includes development of new reagents and techniques for organic synthesis and characterization and for investigation of new organic materials and natural products. Such research provides the basis for designed synthesis of new materials and natural products, and for preparation of compounds important to the chemical and pharmaceutical industries.
- **Experimental Physical Chemistry**—Supports experimental investigations of the physical properties of chemical systems. Scientific issues range from the nature and properties of individual molecules to the behavior of molecules in the aggregate. Areas of current activity include spectroscopy and clusters of molecules, the elastic and inelastic scattering of molecules and photons, the thermodynamics and statistical mechanics of fluids and fluid mixtures, and the kinetics of chemical reactions.
- **Theoretical and Computational Chemistry**—Supports the development of chemical theory, including quantum mechanics, statistical mechanics, and dynamics. This has applications to all areas of chemistry, including support of computer code development.

Much of the Chemistry Division's support of instrumentation and infrastructure is coordinated through the Office of Special Projects. Among these activities is a program of portable Postdoctoral Research Fellowships in Chemistry (see brochure NSF 93-91 for more information); a nationwide network of 60 sites for Research Experiences for Undergraduates (see brochure NSF 93-112 for more information); Research Planning Grants for women and minorities; grants for Faculty Early Career Development; and occasional grants for special purposes in education, curriculum development, and graduate training.

Equally important infrastructure and instrumentation investments, made in the Chemical Instrumentation and Facilities Program (see brochure NSF 93-94 for more information), provide funds to research institutions and consortia thereof for the purchase of multiuser instruments, for major instrumentation development and construction, and for the establishment and support of multiuser research facilities in the chemical sciences. This program is designed to support the following types of academic instrumentation needs: (1) the purchase or upgrade of shared, multiuser instruments (requires a one-third to one-half cost contribution from the grantee institution); (2) the development plans for Chemistry Departments' multiyear instrument centers; (3) instrumentation development, including the construction of new prototype instruments; and (4) the establishment and support of unique national and regional instrumentation facilities. This program focuses on shared instruments and facilities; specialized equipment dedicated for use in particular chemistry research projects is funded as part of individual investigator awards, along with personnel and other direct project costs in other Chemistry Division programs.

Notice that the announcement solicits proposals in major fields of chemical investigation and also suggests related areas of research sponsored by other Divisions of NSF, such as Atmospheric Sciences, Molecular and Cellular Biosciences, Chemical and Transport Systems, Earth Sciences, and Materials Research. As you can see from the following short announcement, Atmospheric Chemistry Research Grants, this program calls for research in a particular area of the chemical sciences.

Atmospheric Chemistry Research Grants ***869***

Research supported includes the measurement and modeling of concentration and distribution of gases and aerosols in the lower and middle atmosphere; chemical reactions among atmospheric species; sources and sinks of important trace gases and aerosols; aqueous phase atmospheric chemistry; transport of gases and aerosols between the troposphere and stratosphere; and improved methods for measuring the concentrations of trace species and their flow through the atmosphere. Submit proposals at any time during the year.
Program No. 47.050
Contact Program Director, Division of Atmospheric Sciences, Directorate for Geosciences, (703) 306-1500

Sponsor
National Science Foundation
4201 Wilson Blvd
Arlington, VA 22230

Other government agencies, such as the Department of Energy and Department of Defense, as well as many private foundations, also offer research opportunities. The following short announcement, Department of Energy Chemical Sciences Research Grants, calls for proposals that relate chemistry to energy resources.

DOE Chemical Sciences Research Grants ***1790***

This program sponsors experimental and theoretical research on liquids, gases, plasmas, and solids, focusing on chemical properties and the interactions of component molecules, atoms, ions, and electrons. The program objective is to expand, through basic research, knowledge in the various areas of chemistry; the long-term goal is to contribute to new or improved processes for developing and using domestic energy resources in an efficient and environmentally sound manner. Disciplinary areas covered include physical, organic, and inorganic chemistry; chemical physics; atomic physics; photochemistry; radiation chemistry; thermodynamics; thermophysics; separations science; analytical chemistry; and actinide chemistry. Applications are accepted at any time.
Program No. 81.049
Contact Director, Chemical Sciences, Basic Energy Sciences, Office of Energy Research, (301) 903-5804

Sponsor
Department of Energy
1000 Independence Ave SW, Forrestal Bldg
Washington, DC 20585

Contacting Program Officers

Once chemists have identified a potential sponsor for their research, they will typically contact the program officer listed on the grant announcement. Program officers are key sources of information about their agency's or foundation's goals and procedures to help researchers shape their proposals. Program officers can tell researchers whether their proposed work falls within program priorities, what issues researchers should address that other applicants may have overlooked, what common mistakes applicants should avoid, and what the criteria for selection and the proposal review process are. In some cases, program officers will review pre-proposals (two- to three-page concept papers) or draft proposals submitted early enough for response and revision.

Talking to program officers can help researchers understand their audience and what will make a proposal persuasive. Learning who will review proposals—whether it is a panel of outside experts in the field, as is the case with NSF peer review, or in-house staff or board members of private foundations—can help proposal writers determine the background of reviewers and the level of technical detail the proposal should include.

After a proposal has been submitted, there may be further negotiations between a researcher and a program officer. In the process of proposal review, questions about the aims, experimental techniques, and budget may arise that need to be addressed before the funding source will approve the proposal and the researcher's work can begin. In other cases, a researcher may not be funded at the time but may, after reading reviews of the proposal, incorporate suggestions and meet criticisms in a revised proposal and resubmit it.

Talking to Past Grantees

Beside program officers, another valuable source of information on proposal writing is past recipients of grant funding. NSF and many private foundations publish the names of grantees and short descriptions of the research funded. One way to get started thinking about writing your own proposal is to talk to a scientist who has submitted a successful grant proposal.

Exercise

Identify a scientist who has received grant funding and is willing to be interviewed. Your instructor can suggest names of scientists on the faculty of your college or university or a nearby one. Use the following questions to prepare for an interview:

1. How did the research you proposed develop out of prior work? What made your proposed research persuasive to reviewers? What do you think proposal reviewers were looking for?

2. How did you identify the funding source you applied to? Did you call or visit the sponsor before writing the proposal? How did you determine there was a good match between the research you proposed and the goals of the funding source? Did the funding source review a pre-proposal or a draft proposal before final submission?

3. What negotiations, if any, took place between you and a program officer? If you revised and resubmitted a proposal, how did you determine what changes to make?
4. What are common mistakes to avoid in preparing a successful proposal?

Prepare a brief oral (five minutes) or written (three pages) report that summarizes what you learned through the interview. Compare what you found with the results of interviews conducted by other members of your class.

WRITING A RESEARCH PROPOSAL

As you have seen, preparing a research proposal involves a number of steps and a good deal of communication. In the following section, we will look at how to write a research proposal and what makes one successful. A research proposal amounts to an argument chemists make to justify particular experimental work. For such arguments to be successful in persuading their readers, effective proposals normally accomplish the following:

- They define a problem that grows out of earlier work in an area of research.
- They explain what makes the problem significant and thereby worthy of investigating.
- They describe an experimental approach that will convincingly address the problem defined.
- They assure readers that the researcher is capable of performing the proposed experiments.
- They anticipate the benefits and significance of the results the research will bring to light.

Writing a research proposal is a good way for you to learn how chemists work—how they define meaningful problems, design experimental approaches to investigate these problems, and anticipate how their research will contribute reliable scientific knowledge to the community of chemists. Writing research proposals, in other words, asks you to do chemistry.

The steps to take are in many ways similar to those experienced chemists follow. It may be helpful for you to think of your instructor's role as similar to the role a program officer plays and to anticipate negotiating the aims and methods of the research you are proposing. In fact, there may be research opportunities for undergraduates on your campus for which you can apply. Your instructor will have more information about such opportunities.

Getting Started: Choosing a Topic

A good topic for a research proposal is one that points to a problem that has emerged in the literature of chemistry and seems to call for more experimental work. Your chemistry instructor may have ideas about possible topics to investigate in your research proposal. You may want to consult with your instructor. Moreover, your instructor may ask for a pre-proposal—a short two- to three-page concept paper—as a way to get started. Here are some possible topics:

- recycling of metals,
- acid rain,
- the ozone layer,
- chemistry of xenon, or
- the recently discovered "fullerenes," C_{60}, C_{70}, and others.

Review the Literature to Define a Problem

Once you have decided on a topic, the next step is to review the literature. The first step in your library research is to identify key articles. Your instructor may give you references to start with. Otherwise, start with the indices of *Chemical Abstracts* as we described in Chapter 6 on pages 136–141."

Read purposefully. The point of your literature search is to identify a problem that indicates the need for further research. As we have mentioned before, scientists pack a great deal of information in a very small amount of space. You may need to read an article several times. Have a textbook handy to look up concepts you are not familiar with and be sure to take notes as you read. Observe carefully what the authors of the article are doing in the article. In particular, when you have reached a point of satisfactory understanding of the paper, answers to the following questions should be clear to you:

1. What is the specific question that the authors have endeavored to answer?
2. How have they proposed to answer this question?
3. How convincing are their results and are any of their results surprising?
4. What aspects of the original question are still unanswered and what are the next logical questions to ask about the topic?

In your literature search, be sure that you reference every article from which you have used information. This will save you time later when you add your references to your research proposal.

Formulate a Problem and Design an Experimental Approach to It

Your literature search, as we have just suggested, is to identify a problem that you can propose to address by an experimental approach. As you consider what you have written, keep in mind that what you want to end up with is a hypothesis regarding a well-defined problem that you can test in a laboratory. This is a good point in your work to consult with your instructor, to get ideas and suggestions about what you might propose.

Plan Your Research Proposal

Like many other kinds of scientific writing, research proposals have a fairly conventional format. Understanding the function of each section of the research proposal—and how each section contributes to a persuasive proposal—can help you write your own research proposal.

In the following sections, we will use as an example a research proposal that seeks to investigate the structure of coal and the reactions of coal to different oxidizing agents.

A research proposal often has the following sections:

- Summary,
- Literature Review,
- Work Proposed,

- Anticipated Results,
- References.

Summary

The summary of a research proposal is normally written after the rest of the proposal is complete. It should start with a short synopsis of the literature survey, then outline very briefly the work that is proposed, and then describe the results expected to be achieved and their significance. In other words, a summary should include in condensed form an overview of what you want to investigate and why. It is helpful for readers to read this section first, so that they can keep in mind the general purpose of the proposed research as they go on to read the more detailed sections that follow. Some readers may only read the summary, so make sure it includes in condensed form the main line of your proposed research.

Here is a sample of an effective summary:

> The structure of coal is complex and heterogeneous. One current model of coal structure describes it as including ordered and disordered domains, some regions that are like tiny crystals of graphite and others that contain a tangle of organic rings and chains. It is reasonable therefore to infer that different parts of coal structure will undergo different chemical reactions. To investigate this inference, I will react powdered coals with a variety of different oxidizing agents such as sodium hypochlorite and hydrogen peroxide. The extent of reaction in each case will be determined by the change of mass of the coal. Correlation of the results with the known reactions of the oxidizing agents should give information on the nature of the coal structure.

Notice that this summary doesn't explain fully what work has already been done in the field and what research is proposed. Instead it focuses briefly on a particular problem—the heterogeneous structure of coal—and suggests, again briefly, an experimental approach to investigate the problem. Finally, again briefly, it indicates the potential significance of the results of the research proposed. In other words, this summary gives an overview.

Many issues, of course, remain to be developed to make this a persuasive proposal, issues that will be treated in the literature review and work proposed sections. What coal is and what its different regions are will need to be explained more fully. The writer will have to tell what evidence has been given to support the model of the coal structure that is being assumed. Reactions that have already been done will have to be described to show that the proposed work is feasible, as well as demonstrate that it will not be repeating work that has already been performed. Key terms will have to be defined or referenced as they appear.

The Literature Review

The function of the literature review section is to explain more fully how the particular problem the researcher proposes to investigate and the experimental approach the researcher proposes to undertake emerge from earlier work. The purpose of this section is not simply to give background information. Rather it is to formulate a problem that can be addressed by experimental means.

Notice in the following passages how the writer has effectively developed a persuasive case for her research. The writer begins with accepted concepts and definitions:

> A complex and heterogeneous material, coal is not a single material but includes a range of different compositions from the very soft, brown lignites to the hard, intensely black and reflective anthracites. The harder coals are said to have higher *rank,* a parameter of coal that increases with the amount of time that it is exposed to the heat and pressure of underground burial. A model that is widely but not universally accepted of coal includes ordered and disordered domains. . . .

Next, the literature survey presents evidence and references for the model of the coal structure that is being used:

```
    X-ray diffraction patterns of coal¹ show the
principal diffraction peak of graphite. Graphite
itself has a layered structure composed of sheets
of carbon atoms. The coal diffraction increases in
intensity as coal rank increases. A reasonable in-
terpretation of this observation is that the coal
structure includes small graphite-like crystals.
However, the diffraction in coal is broader than
in pure graphite crystals and coal lacks the com-
plete three-dimensional structure of graphite. The
broadness of the diffraction indicates that the
crystals are extremely small² with each one con-
sisting of only a few carbon sheets.³ The remain-
der of the structure appears not to have crystal-
lographic order and is most likely a highly
disordered array of various organic links and
chains. . . .
```

Notice that this excerpt from the Literature Review is detailed, thorough, and carefully referenced. It is important for readers to see that the previous work has been well understood and that the new research grows out of what has already been done. Notice too that the writer has been careful not to overstate the conclusions of other work. For example, she says, "a reasonable interpretation of this observation is that . . . ," rather than stating that the structure is known. Nature is complex, and it takes a great deal of work to understand the structure of the physical world. Besides, the writer does not want to give the impression that, contrary to fact, the problems she wishes to attack are well understood.

Work Proposed

After covering the details of coal chemistry thoroughly, the stage is set for the Work Proposed section. This is really the heart and soul of the proposal. In the Literature Review, the writer showed that she was familiar with the previous work on her subject. In the Work Proposed section she must demonstrate that she has a good research idea, explain what she is going to do, and convince her readers that her plan will work.

If you are preparing this proposal as an assignment, your instructor will explain the level of detail he or she expects. In some assignments the emphasis will be on formulating reasonable questions and you will not be expected to give the details of the actual experiments to be performed. In other cases, presentation of these experimental details will be considered crucial.

The Work Proposed section begins by explaining the problems that need to be solved and then stating the basic premise of the proposal—in this case that chemical reactions will react to different extents within the ordered and disordered domains of the coal structure.

The structure of coal is still largely a mystery because none of the techniques that the chemist normally employs to probe its chemical structure work very well. The reasons for this are that coal is not totally soluble, it is opaque, and it can't be vaporized. These characteristics prevent coal from being studied by some techniques and greatly hamper the effectiveness of the others. Therefore relatively small steps toward the understanding of this important but enigmatic material may be as critical as larger steps toward understanding other compounds.

The large structural difference between the ordered and disordered domains of the coal structure makes it likely that these two domains will react very differently with chemical reagents. In general, we would expect the disordered domains to be highly reactive to most reagents, particularly oxidizing agents, and the graphite-like crystals to be much more resistant to reaction.

Note how this statement stresses how little is known about the coal structure, thus accenting the importance of the work proposed. Then the basic premise of the proposal is given.

Next it is necessary to explain the proposed chemical reactions.

Coal has been shown by infrared spectroscopy[4] to have a number of functional groups that are susceptible to oxidation. These include sulfur groups such as sulfide and mercaptan, oxygen groups such as carbonyl, and ethylene carbons. All of these groups are well known to react with a wide range of oxidizing agents. On the other hand, if coal does indeed contain small, graphite-like crystals, their susceptibility to oxidizing agents should be much less. The model compound for these small crystals is graphite with its many carbon sheets extending for huge numbers of carbon atoms. Graphite is noted for being extremely unreactive and often serves as a low-cost inert material in uses such as batteries. Thus it would seem that the graphite-like crystals would be unreactive as well, although we must bear in mind that the high surface-to-area ratio of a tiny graphite-like crystal compared to the low ratio for a graphite crystal makes the tiny crystal at least somewhat more reactive.

Again the writer is thorough but careful. She doesn't state that the tiny crystals in the coal will react exactly like graphite because, as she says, the sizes are so different. It is important not to make an overstatement that will give a reader the notion that the proposal is based on faulty assumptions.

The writer continues by developing the logic of the proposal.

Oxidation reactions should supplement other techniques of structure determination by providing a probe into the coal structure. These chemical oxidations could be used to explore whether coal actually has the ordered and disordered domains. These reactions could be accompanied by X-ray studies. For example, if part of the disordered region were oxidized away, this should be detectable in the X-ray pattern by a strengthening of the principal graphite diffraction peak.

Then she presents the details of the proposed experiments.

```
   Coals of a variety of ranks from lignites to
anthracites will be powdered and screened so that
the grains are below a common size. The powders
will be placed in a vacuum desiccator for several
days to remove adsorbed gases from the surface.
Weighed samples will be placed in a flask and re-
acted with an aqueous oxidizing agent with stir-
ring and under nitrogen. These reactions will be
performed at set temperatures and times. Following
reaction the samples will be filtered, rinsed, and
dried in a vacuum desiccator under the same condi-
tions that were used for the unreacted sample. . .
```

In its entirety, the experimental section is an accurate and quite detailed description of how the initial experiments will be done. Don't leave out any crucial details at this stage such as running the reaction under nitrogen or drying a product. Of course, once the work is started it is expected that the experimental procedure will vary as the project develops.

Once a thorough Work Proposed section has been written, the nature of the results can be anticipated.

Anticipated Results

Anticipated Results is the final section of the research proposal. Because you do not yet have any results, this section is likely to be fairly brief. However, it is an important one. For your proposal to be persuasive, the results you anticipate should appear capable, at least in theory, of making a contribution to the problem you are addressing. An excerpt of this section of the above proposal is:

```
The difference in masses of the unreacted and re-
acted coal samples should give a measure of the
fraction of the coal structure that is disordered.
The size of the disordered fraction should de-
crease as the rank of the coal decreases and this
would serve as a test of the method. Correlation
with X-ray diffraction results could be tested and
reacted coals should show greater diffraction from
ordered crystals.
```

If these tests are in agreement with this proposal, strong proof supporting the ordered-disordered model of coal will have been gained. Disagreement about the nature of the coal structure is still strong and lack of understanding of this structure is hampering progress in studying coal reactivity and effective use of coal as a fuel and a basic source of chemicals.

The Anticipated Results section is difficult. Say as much as you can about what you expect to happen but don't go too far out on a limb and give readers a chance to say the proposal is unrealistic.

References

References should be listed in a separate section at the end of the proposal. If a reference appears twice in the body of the proposal, use its original number each time. Do not have the same reference appear twice in your list. The format for references is given in Chapter 6, pages 129–133. The references for this proposal are these:

1. Gerstein, B. C.; Murphy, P. D.; Ryan, L. M. In *Coal Structure;* Meyers, R. A., Ed.; Academic: New York, 1982; pp. 87-129.
2. Warren, B. E. *Phys. Rev.* **1941,** *59,* 693-698.
3. Diamond, R. *Acta Crystallogr.* **1957,** *10,* 359-364.
4. Tshamler, H.; de Ruiter, E. In *Chemistry of Coal Utilization;* Lowry, H. H., Ed.; Wiley: New York, 1963.

Drafting

Once you have sketched out your material, it is time to write the first draft. Remember that it makes the most sense to write the Summary last, after the other sections, to make sure it gives a proper overview of what you are proposing and why. Use what you know about the function of each section in a research proposal to help you organize your material. Your best bet is to write the first draft fairly quickly. This will allow you to see your writing on the page and to make the necessary reorganization when you revise. Most of all, don't expect to write a "perfect" draft the first time around. Give yourself the opportunity to see your work and make needed changes when you revise.

Revision

Now that you have a draft in front of you, you can fine-tune each section so that you end up with the most persuasive proposal possible. You may find that you need to move material from one section to another or add or delete material. In any case, here are some questions you can ask yourself—or ask a classmate or friend to answer—to give you ideas about what revisions are needed.

- Does the Summary give a well-focused but brief overview of the problem you're investigating, the experimental approach to the problem, and the significance of anticipated results? As you read over this section, the trick will be determining what is too much and what is too little. Do you need to add or delete material?
- Does the Literature Review explain clearly how the problem you have defined is a meaningful one that emerges from earlier work? Do you explain and reference the concepts and models you are using in the proposed research?
- Does the Work Proposed section indicate clearly what research you are proposing and how this research will enable you to answer the question you are addressing?
- Does the Anticipated Results section explain how the results you hope to get will enable you to answer the questions you are investigating? Do you explain the significance of these results to work in the area of study?

Editing and Proofreading

Make sure you give yourself enough time to do a thorough editing of your final draft. At this point, look especially for sentences or phrases that are hard to follow. Ask a classmate or friend to read your final draft and point to passages that are not as clear as they could be. Finally, do a careful proofreading of your proposal, checking for misspellings, grammatical errors, or usage problems.

8

Chemistry and the Public: Writing to Inform and Persuade

Most scientists are called on from time to time to explain technical information to the public. A quick look at recent news reveals a variety of occasions that call on scientists to address scientific issues for broad audiences—whether the issue is acid rain, the safety of nuclear power plants, or the greenhouse effect. In addition, there is a considerable need in both the public and private sectors for writers who understand science and technogy and can explain it to lay people. Government agencies, environmental organizations, health care facilities, and private corporations often employ technical and scientific writers to communicate to the public through news releases, articles, brochures, leaflets, and manuals. Many newspapers have one or more science writers on the staff who contribute regular articles and news reports.

As you will see in this chapter, writing about science for the public differs in a number of important respects from writing for the community of scientists. We will be looking in particular at how science writers report on recent scientific news and findings and how science writers take positions on controversial issues in order to influence public policy.

REPORTING SCIENCE: WRITING TO INFORM

The purposes of writing about science for the public will vary depending on the science writer's situation and the effects he or she wants to have on the audience. In the case of science writers who report on scientific news and recent findings for newspapers, the purpose of reporting is to *inform* the public. Informative science writing seeks largely to explain events and concepts to readers, to help readers stay knowledgeable and up to date about current developments in science.

Informing readers about science, of course, is not simply a matter of presenting the facts. For one thing, science writers need to consider the audience that will be reading their work and have a fairly good idea of what that audience is likely to know already about the science they're reporting on. In this regard, science writers will typically make some assumptions about their readers and construct in their own minds a working sense of who their audience is and what the the audience will know and understand.

Second, not only do science writers need to be accurate about the information they are reporting, but they also need to find writing strategies that explain the information and its significance to as many readers as possible. To do this, writers usually try to establish the focus of the article clearly and comprehensibly, to define key terms and concepts, and to explain important processes in accessible ways. In other words, science writers want to make the science they're writing about meaningful to their readers.

A good source of science writing to inform the public is the "Science Times" section that appears every Tuesday in the *New York Times.* The following article is taken from that section. Read it carefully and evaluate how successful it is.

Chemists Dissect the Colors of Flowers

What makes geraniums red, hyacinths purple and day flowers blue has been debated among chemists for 70 years. While they agree that a few chemicals in flowers cause myriad hues in the petals, scientists have differed on how these colors develop.

Now researchers say they have determined the structure of anthocyanins, one type of floral pigment. Dr. Kumi Yoshida, an assistant professor of organic chemistry at Sugiyama Jogakuen University in Japan, who is an author of the report in the current issue of the journal *Nature*, said anthocyanin was responsible for the day flower's blue color.

The anthocyanin in the day flower, *commelina communis*, is an unstable pigment complex with a daisy-like configuration of molecules extending from a positively charged central atom. In the past, the chemical structures of anthocyanins could not be studied because they would fade and break down when analyzed. The researchers stabilized the pigment complex by building the network from its isolated components and replacing the central magnesium ions with similarly charged cadmium atoms. That did not alter the chemical association but created an array stable enough to be tested outside the flower.

Using X-ray crystallography and nuclear magnetic resonance, the research team, led by Dr. Tadao Kondo, an assistant professor of organic chemistry at Nagoya University in Japan, concluded that the development of different hues depends on interactions between atoms of the pigment as well as interactions between anthocyanin and other pigments within the petal.

Writing Assignments

1. Analyze this article in terms of the following questions.
   a. Who is the intended audience of the article? How can you tell?
   b. How does the article establish its focus? How likely (or not) will this focus help readers make sense of the information that follows?
   c. How does the article describe the actual science involved? What does the writer explain? What does the writer seem to assume does not need to be explained?
   d. In your view, how successful has the writer been in explaining the science and its significance to a broad audience of readers?
2. Rewrite the article in a short form that would be suitable for an audience of sixth-grade students.
3. Choose one of the following topics or pick a topic of your own to write an informative article on the subject suitable for the audience of the "Science Times" section of the *New York Times*.
   - Scanning tunneling microscopy (see articles, *Nature*, **344,** 524 [1990] and *Nature*, **346,** 294 [1990]).
   - How silicon chips for electronics are made.
   - How magnesium metal is extracted from sea water.
   - Synthetic diamonds (see the article, *Science*, **254,** 653 [1991]).

- The recent Nobel prize for neutron diffraction.
- What are liquid crystals?
- Superlight, superstrong materials now used in bicycle frames, car bodies, and rockets.
- Why straight hair can be curled and curly hair straightened.
- Recent developments in ceramics.
- Brand name and generic drugs: how do you get the best bargain?
- How are colors achieved in fireworks, glass and pottery, and fabrics?
- How do soaps and detergents work and how do you choose the best to use?

CHEMISTRY AND "CHEMICALS"

Chemistry gets a good bit of criticism these days. Generally this criticism is directed toward "chemicals"—chemicals in our air, chemicals in our water, chemicals in our food. And chemists get blamed for the unwelcome presence of these "chemicals."

But what are "chemicals?" "Chemicals" is not a word that chemists actually use much. Chemists talk about elements and compounds; together these are known as pure substances. And they talk about mixtures, which are compositions of matter containing more than one substance. These terms—element, compound, substance, and mixture—have very specific meanings, and they describe every single bit of matter in the universe. So where does the word "chemicals" fit in?

There certainly is a chemical industry. Among the famous names in the chemical industry in the United States are DuPont, Union Carbide, and American Cyanamid. The chemical industry is, by the way, one of only two industries in the United States for which the value of exports exceeds the value of imports. All of these chemical companies now have a wide range of activities. However, the chemical part of their businesses involves producing and selling elements, compounds, and mixtures. Some of these go directly to the consumer as cleaning aids, insect repellents, gardening supplies, and other products. But most of the products of the large chemical companies are sold to industries for making glass, paint, plastics, paper, electronic components, preservatives, fertilizers, and countless other products including essentially all of the items that we use and experience in our lives.

We could define chemicals as the products of chemical companies. But this doesn't really help us much to understand what "chemicals" are. Take ethyl alcohol, for example, which is often produced by chemical companies but is also produced by yeast in the bread baking process. Is ethyl alcohol a chemical in the first case but not in the second? From a scientific point of view, a substance is not characterized by how it was made but by what it is. We know that all molecules of ethyl alcohol or of any other compound are identical, containing the exact same atoms. Thus, by any reasonable definition, a "chemical" must be a compound, an element, or a mixture. Everything is made up of chemical elements and compounds. Therefore, everything is a chemical.

It is obviously not clear to the public exactly what chemicals are. Nonetheless, there can be little question that the public is suspicious of them. The following example involves a very useful chemical compound in the kitchen. Vanilla beans contain a compound called vanillin, which has a particular smell and a taste that give vanilla its special character as a flavoring. The vanillin can be extracted from vanilla beans to flavor ice cream, cakes, and many other delicious confections. (The extraction process, incidentally, involves commercial solvents obtained from chemical companies.) Alternatively, vanillin itself can be produced by a chemical company from other compounds and sold as "artificial" vanilla. The vanillin in artificial vanilla is absolutely identical to that in the natural vanilla extract. That is, it contains exactly the same atoms arranged in exactly the same way. Furthermore, the synthetic vanillin is purer. It has undergone an extensive purification process and does not contain the impurities introduced in the picking, transportation, storing, and extraction of the beans. And artificial vanilla is a lot cheaper. Take a look at the prices the next time you are in the grocery store. However, most people would prefer the natural vanilla; is there any good reason for this?

Public attitudes toward "chemicals" and "natural" items are often based not on scientific understanding but rather on a widespread suspicion of science. Deodorants that are sold that are "chemical free." What can they possibly contain? Tom's Toothpaste from Maine tells us that the fluoride compound that it contains is made from naturally occurring calcium fluoride. Of course it's made from naturally occurring calcium fluoride. All compounds containing fluorine, from Teflon to hydrofluoric acid for etching glass, are made from naturally occurring calcium fluoride, the only significant source of the element. A city councilor in Worcester, Massachusetts, claims to be allergic to "all chemicals, natural and synthetic." She must find her life very restricted. Examples like this abound, and the claim that a product is "natural" or "chemical free" is a common one.

Writing Assignments

1. Articles in newspapers and magazines containing incorrect chemistry are very common. "Trailer truck overturns exposing thousands to deadly nitrogen gas." Find a recent article with incorrect chemistry and explain why it is incorrect.
2. Write a few paragraphs designed to convince people that table salt, ordinary NaCl, is a highly dangerous compound. Use what you know about the reactions that can be used to prepare NaCl.
3. Find an article in a newspaper or magazine in which the author is using the public's suspicion of chemicals to advance a particular opinion. Explain how this is being done in the article.

CHEMISTRY AND PUBLIC OPINION: WRITING TO PERSUADE

Earlier, we considered writing for the public with the purpose of *informing.* In this section, we will look at situations in which the writer's purpose is not simply to inform readers but also to *persuade* them. Scientists testifying before a congressional committee about needed emission standards for automobiles, environmentalists advocating recycling programs, or citizens trying to close down a toxic waste dump are all seeking to persuade their audience to a particular point of view and a particular course of action.

To persuade their readers, writers need to analyze their audience carefully. Writers will typically try to find points of agreement and values they share with their readers so that they can develop their position in terms that others are likely to find amenable. For example, the desire for personal health, clean air and water, safe food, and sustainable resources is often a starting point in advocating particular kinds of public policy. But the suspicion directed toward "chemicals," which we discussed in the last section, is also often used to influence public opinion by mobilizing people's fears, whether or not they are realistic and scientifically grounded.

Second, writers who seek to persuade their readers need to identify and define clearly the pressing issue that seems to call for some form of action. Here writers will typically inform their readers about, say, the extent of toxic dumping and its potentially disastrous effects as background to their advocacy of new regulations. Informing readers is the first step in persuading them. If readers do not acknowledge that a problem exists, they will be unlikely to want to do anything about it.

Finally, writers who seek to persuade their readers need to offer good reasons and, at times, to refute other positions or interpretations. Besides defining a problem or issue as one that requires attention, writers will typically explain to readers how and why the negative effects of, say, acid rain, or the beneficial possibilities of recycling seem to necessitate a particular course of action. Skillful writers, moreover, do not pretend that theirs is the only available position on an issue or interpretation of a problem. Their writing will be more persuasive to readers if they can demonstrate how they differ from others and why.

The following readings illustrate how science writers have used persuasive writing to advance their claims and positions.

Insecticides

Since all substances are chemicals, it is reasonable to suspect that some uses of chemicals are undesirable. An example is the unrestrained use of insecticides following the brilliant successes of controlling disease and human discomfort when insecticides were first introduced. In 1962 Rachel Carson published *Silent Spring*, a warning to a society that, at the time, was inclined to overlook the growing evidence that chlorinated hydrocarbons used in insecticides such as DDT were having dangerous consequences. When her book appeared, it was attacked quite violently by the chemical industry and by some scientific organizations. It is now considered a classic of the environmental movement.

As you read the following excerpt from *Silent Spring*, think about what Carson is trying to accomplish with her writing and how she is doing it. Also consider the scientific content of the work and how she uses facts and perceptions about science.

Silent Spring

Rachel Carson

Elixirs of Death

For the first time in the history of the world, every human being is now subjected to contact with dangerous chemicals, from the moment of conception until death. In the less than two decades of their use, the synthetic pesticides have been so thoroughly distributed throughout the animate and inanimate world that they occur virtually everywhere. They have been recovered from most of the major river systems and even from streams of groundwater flowing unseen through the earth. Residues of these chemicals linger in soil to which they may have been applied a dozen years before. They have entered and lodged in the bodies of fish, birds, reptiles, and domestic and wild animals so universally that scientists carrying on animal experiments find it almost impossible to locate subjects free from such contamination. They have been found in fish in remote mountain lakes, in earthworms burrowing in soil, in the eggs of birds—and in man himself. For these chemicals are now stored in the bodies of the vast majority of human beings, regardless of age. They occur in the mother's milk, and probably in the tissues of the unborn child.

All this has come about because of the sudden rise and prodigious growth of an industry for the production of man-made or synthetic chemicals with insecticidal properties. This industry is a child of the Second World War. In the course of developing agents of chemical warfare, some of the chemicals created in the laboratory were found to be lethal to insects. The discovery did not come by chance: insects were widely used to test chemicals as agents of death for man.

The result has been a seemingly endless stream of synthetic insecticides. In being man-made—by ingenious laboratory manipulation of the molecules, substituting atoms, altering their arrangement—they differ sharply from the simpler insecticides of prewar days. These were derived from naturally occurring minerals and plant products—compounds of arsenic, copper, lead, manganese, zinc, and other minerals, pyrethrum from the dried flowers of chrysanthemums, nicotine sulphate from some of the relatives of tobacco, and rotenone from leguminous plants of the East Indies.

What sets the new synthetic insecticides apart is their enormous biological potency. They have immense power not merely to poison but to enter into the most vital processes of the body and change them in sinister and often deadly ways. Thus, as we shall see, they destroy the very enzymes whose function is to protect the body from harm, they block the oxidation processes from which the body receives its energy, they prevent the normal functioning of various organs, and they may initiate in certain cells the slow and irreversible change that leads to malignancy.

Yet new and more deadly chemicals are added to the list each year and new uses are devised so that contact with these materials has become practically worldwide. The production of synthetic pesticides in the United States soared from 124,259,000 pounds in 1947 to 637,666,000 pounds in 1960—more than a fivefold increase. The wholesale value of these products was well over a quarter of a billion dollars. But in the plans and hopes of the industry this enormous production is only a beginning.

A Who's Who of pesticides is therefore of concern to us all. If we are going to live so intimately with these chemicals—eating and drinking them, taking them into the very marrow of our bones—we had better know something about their nature and their power.

Although the Second World War marked a turning away from inorganic chemicals as pesticides into the wonder world of the carbon molecule, a few of the old materials persist. Chief among these is arsenic, which is still the basic ingredient in a variety of weed and insect killers. Arsenic is a highly toxic mineral occurring widely in association with the ores of various metals, and in very small amounts in volcanoes, in the sea, and in spring water. Its relations to man are varied and historic. Since many of its compounds are tasteless, it has been a favorite agent of homicide from long before the time of the Borgias to the present. Arsenic is present in English chimney soot and along with certain aromatic hydrocarbons is considered responsible for the carcinogenic (or cancer-causing) action of the soot, which was recognized nearly two centuries ago by an English physician. Epidemics of chronic arsenical poisoning involving whole populations over long periods are on record. Arsenic-contaminated environments have also caused sickness and death among horses, cows, goats, pigs, deer, fishes, and bees; despite this record arsenical sprays and dusts are widely used. In the arsenic-sprayed cotton country of southern United States beekeeping as an industry has nearly died out. Farmers using arsenic dusts over long periods have been afflicted with chronic poisoning; livestock have been poisoned by crop sprays or weed killers containing arsenic. Drifting arsenic dusts from blueberry lands have spread over neighboring farms, contaminating streams, fatally poisoning bees and cows, and causing human illness. "It is scarcely possible . . . to handle arsenicals with more utter disregard of the general health than that which has been practiced in our country in recent years," said Dr. W. C. Hueper, of the National Cancer Institute, an authority on environmental cancer. "Anyone who has watched the dusters and sprayers of arsenical insecticides at work must have been impressed by the almost supreme carelessness with which the poisonous substances are dispensed."

Modern insecticides are still more deadly. The vast majority fall into one of two large groups of chemicals. One, represented by DDT, is known as the "chlorinated hydrocarbons." The other group consists of the organic phosphorus insecticides, and is represented by the reasonably familiar malathion and parathion. All have one thing in common. As mentioned above, they are built on a basis of carbon atoms, which are also the indispensable building blocks of the living world, and thus classed as "organic." To understand them, we must see of what they are made, and how, although linked with the basic chemistry of all life, they lend themselves to the modifications which make them agents of death.

The basic element, carbon, is one whose atoms have an almost infinite capacity for uniting with each other in chains and rings and various other configurations, and for becoming linked with atoms of other substances. Indeed, the incredible diversity of living creatures from bacteria to the great blue whale is largely due to this capacity of carbon. The complex protein molecule has the carbon atom as its basis, as have molecules of fat, carbohydrates, enzymes, and vitamins. So, too, have enormous numbers of nonliving things, for carbon is not necessarily a symbol of life.

Some organic compounds are simply combinations of carbon and hydrogen. The simplest of these is methane, or marsh gas, formed in nature by the bacterial decomposition of organic matter under water. Mixed with air in proper proportions, methane becomes the dreaded "fire damp" of coal mines. Its structure is beautifully simple, consisting of one carbon atom to which four hydrogen atoms have become attached:

```
H     H
 \   /
   C
 /   \
H     H
```

Chemists have discovered that it is possible to detach one or all of the hydrogen atoms and substitute other elements. For example, by substituting one atom of chlorine for one of hydrogen we produce methyl chloride:

```
H     Cl
 \   /
   C
 /   \
H     H
```

Take away three hydrogen atoms and substitute chlorine and we have the anesthetic chloroform:

```
H      Cl
 \   /
   C
 /   \
Cl     Cl
```

Substitute chlorine atoms for all of the hydrogen atoms and the result is carbon tetrachloride, the familiar cleaning fluid:

```
Cl     Cl
  \   /
    C
  /   \
Cl     Cl
```

In the simplest possible terms, these changes rung upon the basic molecule of methane illustrate what a chlorinated hydrocarbon is. But this illustration gives little hint of the true complexity of the chemical world of the hydrocarbons, or of the manipulations by which the organic chemist creates his infinitely varied materials. For instead of the simple methane molecule with its single carbon atom, he may work with hydrocarbon molecules consisting of many carbon atoms, arranged in rings or chains, with side chains or branches, holding to themselves with chemical bonds not merely simple atoms of hydrogen or chlorine but also a wide variety of chemical groups. By seemingly slight changes the whole character of the substance is changed; for example, not only what is attached but the place of attachment to the carbon atom is highly important. Such ingenious manipulations have produced a battery of poisons of truly extraordinary power.

DDT (short for dichloro-diphenyl-trichloro-ethane) was first synthesized by a German chemist in 1874, but its properties as an insecticide were not discovered until 1939. Almost immediately DDT was hailed as a means of stamping out insect-borne disease and winning the farmers' war against crop destroyers overnight. The discoverer, Paul Müller of Switzerland, won the Nobel Prize.

DDT is now so universally used that in most minds the product takes on the harmless aspect of the familiar. Perhaps the myth of the harmlessness of DDT rests on the fact that one of its first uses was the wartime dusting of many thousands of soldiers, refugees, and prisoners, to combat lice. It is widely believed that since so many people came into extremely intimate contct with DDT and suffered no immediate ill effects the chemical must certainly be innocent of harm. This understandable misconception arises from the fact that—unlike other chlorinated hydrocarbons—DDT *in powder form* is not readily absorbed through the skin. Dissolved in oil, as it usually is, DDT is definitely toxic. If swallowed, it is absorbed slowly through the digestive tract; it may also be absorbed through the lungs. Once it has entered the body it is stored largely in organs rich in fatty substances (because DDT itself is fat-soluble) such as the adrenal, testes, or thyroid. Relatively large amounts are deposited in the liver, kidneys, and the fat of the large, protective mesenteries that enfold the intestines.

Writing Assignments

1. Write a short one-paragraph summary. Compare your summary to those of your classmates. Don't argue about which is better. Instead, try to explain how and why they differ or are alike.

2. What effect do you think Carson is trying to create in her opening sentence, "For the first time in the history of the

world, every human being is now subjected to contact with dangerous chemicals, from the moment of conception until death"? What evidence does she present to back up this statement? What feelings do you think she is trying to tap in the reader?

3. Write a short essay that analyzes the techniques Carson uses to persuade the reader to her point of view. In particular, consider how she uses chemical structural formulas. What does this tell you about her assumed audience? What kind of credibility does it lend to Carson's writing? Give your own overall evaluation of Carson's techniques and writing style.

Acid Rain

Another environmental problem that has caused a lot of controversy is "acid rain." Often the debate about acid rain is posed as a matter of saving the environment versus saving jobs. The deleterious effects of acid rain are felt by forests and cropland, fresh water lakes, public buildings and monuments, and human health from the metal contamination of acid water when passed through metal pipes. The damage of acid rain to economic activity in the northeastern U.S. is estimated to cost from $5 billion to as much as $20 billion per year. Control of sulfur oxide emissions by Midwestern power sources is estimated to have a price tag of $7 to $18 billion and result in about 50,000 lost jobs in the high-sulfur coal-producing states in the band between West Virginia and Illinois. As you will see, the point of view in the following reading selection by Robert Ostmann tends to favor lowering the use of energy in general and the consumption of coal in particular.

Acid Rain: A Plague Upon the Waters

Robert Ostmann, Jr.

The rain has turned to acid

Sulfuric showers. April acid. Nitric rainbow.

These are nasty, toxic, corrosive laboratory names for what was once the universal symbol of natural purity

We spent a decade and more in the 1960s and 1970s calling for a halt to the ever-increasing pollution of the planet, but we somehow forgot about the rain. We turn a vigilant eye to pipes disgorging vile fluids and to smokestacks soiling neighborhoods with clouds of noxious emissions. But when the air appeared cleaner, most of us looked no farther. We just assumed that of all things, the rain would remain pure. We did not realize until the raindrops were already sour that the destruction had begun; that if you pump great volumes of poison into the complex global organism of the environment, the ill effects will spread far beyond the point of injection.

After the first alarms from scientists—largely ignored—led to more and louder alerts, we finally began to refocus our sights. What we found is that the poision had indeed spread, insidiously, on a scale we had never imagined. We found that both the rain and the snow actually had become dilute sulfuric, nitric, and in some cases hydrochloric acid. The transformation is the result of a complicated, and not yet fully deciphered, atmospheric recipe whose key ingredients are sun, wind, water, and chemical pollutants. These pollutants—mainly sulfur dioxide and nitrogen oxides—are released into the atmosphere wherever coal and, to a lesser degree, oil or natural gas are burned: from the smokestacks of electric generating plants, metal smelters and industrial boilers, and from the exhaust pipes of motor vehicles. The same pollutants also arise from natural sources such as volcanic eruptions, forest fires, and the slow bacterial decomposition of organic matter. The rapidly increasing portion contributed by humans, however, has caused the trouble. In one year the sulfur dioxide emissions from a large coal-fired power plant can equal the huge amount released by the May 18, 1980, eruption of Mount St. Helens.

Each year human activity injects at least 100 million metric tons of sulfur dioxide and 35 million metric tons of nitrogen oxides into an already polluted atmosphere.[3] Americans contribute some 30 million metric tons of sulfur dioxide and about 26 million metric tons of nitrogen oxides to the acid-forming pollutants over North America.[4] The Canadians, primarily their metal smelting industry, release another 5 million metric tons of sulfur dioxide and 2 million metric tons of nitrogen oxides.[5] Over the rest of the Northern Hemisphere, the European nations are responsible for more than 50 million metric tons of sulfur dioxide and 4 million metric tons of nitrogen compounds.[6] Then there are the Russians, whose current emissions are unknown, and, increasingly, the Chinese, who with sulfur dioxide emissions of 15 million metric tons, are rushing headlong into a highly industrialized, acidic future.[7]

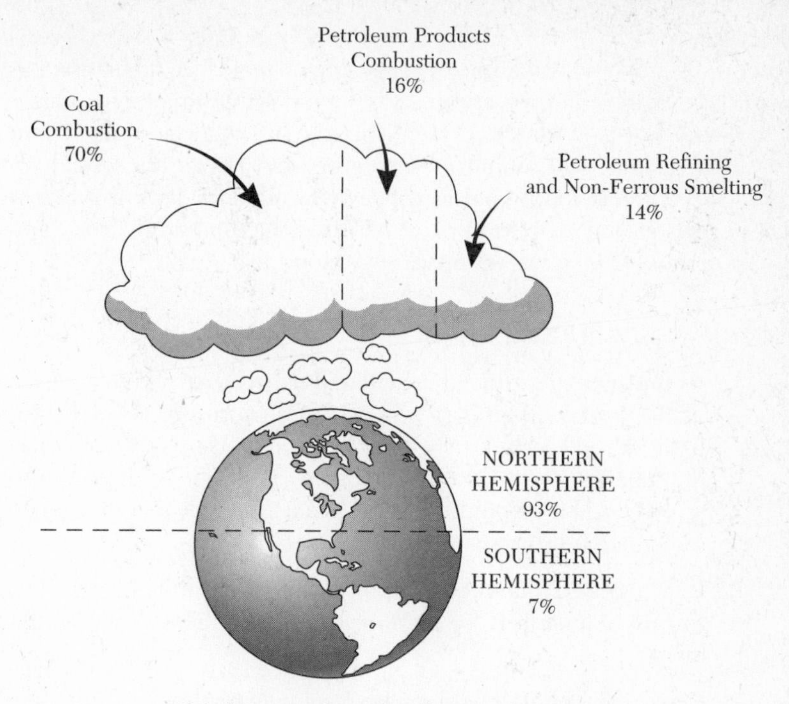

Global Human-Caused Sulfer Dioxide Emissions

The Earth's weather systems weave together these innumerable pollution plumes into great regional masses of contaminated atmosphere that mix, swirl, and carry the sulfur and nitrogen compounds hundreds, and sometimes thousands, of miles from where they were originally discharged into the air. Scientists now believe that from one-half to two-thirds of the compounds that enter these masses of atmospheric pollution react with moisture and other particles in the air and are transformed into molecules of dilute sulfuric and nitric acid that condense and fall to earth as acid rain or snow. In areas that receive little rainfall, most acid-forming pollution reaches the earth as dry particles—a phenomenon known as dry deposition—or combines with water particles to form acid dew or mist. Southern California has its own special brand of acid smog.

Taken together in all of its forms, wet and dry, acid deposition has emerged in recent years as one of the most serious environmental threats to life on our planet. In fact, its visible and devastating ecological effects have caused scientists and governments around the world to sound cries of alarm. In the scientific community today, there is general—though not unanimous—agreement that acid deposition has begun to extinguish entire species of fish and other aquatic life forms in vast areas of the Northern Hemisphere. In addition, many scientists believe that continuing or worsening acid deposition could reduce the productivity of vital forests and farmlands, disrupt the crucial, life-sustaining process of plant photosynthesis in large areas, and poison some drinking water supplies and food fish stocks.

Writing Assignments

1. In this excerpt, notice particularly Ostmann's use of chemical names and amounts, as well as the illustration, "Global Human-Caused Sulfur Dioxide Emissions." Write a short essay that analyzes how Ostmann uses this scientific information to develop his own point of view. What does his use of science tell you about his assumed audience?

2. How does Ostmann take advantage of the public's suspicions toward science? How do you feel about the techniques he uses?

3. Find an account of acid rain in a science textbook. There may well be a discussion in your chemistry textbook (or consult William L. Masterson and Cecile N. Hurley, *Chemistry: Principles and Reactions*, Saunders College Publishing). Write a short essay that compares the textbook's account of acid rain to Ostmann's. Do they present the same scientific information? Are their interpretations the same or do they differ? How are they alike or different in tone and style?

Dioxin

Another compound that has received enormous public attention in recent years is 2,3,6,7-tetrachloro dibenzo dioxin, popularly known simply as dioxin. This compound is formed as a byproduct in the preparation of the herbicide or weed killer, 2,4,5-T, a component of Agent Orange, a defoliant used in the Vietnam War. It also appears to be formed in the combustion of some chlorine-containing organic compounds and plastics. It causes a serious skin disorder called chloracne and has been implicated as a cause of cancer and birth defects.

Dioxin, Agent Orange: The Facts

Michael Gough

Dioxin lurks seemingly everywhere, as though spread around the country by some malevolent force. Undetectable to the human senses, dioxin has been found by use of sophisticated scientific apparatus in expected places, like old chemical plants and chemical dumps, and unexpected places, like people's front yards, pristine lakes in virgin forests, and Boy Scout encampments. Each discovery sets off a wave of headlines and controversies about human health. In newspapers and news programs, experts from universities and public interest groups highlight predictions of dire tragedies, bracketing the word "dioxin" with "cancer-causing chemical" and "most toxic chemical known." In the same article or newscast, experts from chemical companies and various levels of government generally respond with unsuccessful reassurances that the risks are exaggerated. In some cases, the soothing words are belied by the government's providing elaborate protective devices to the workers it sends to sample the same area where it had downplayed the risks. As days pass, the news value of the story declines, and the latest dioxin story fades away, perhaps to oblivion, perhaps to later resurrection as a reminder of the prevalence of the dioxin risk when yet another discovery is made. The story has no resolution; it just goes away. . . .

Unknown numbers of people came into contact with dioxin when dioxin-containing herbicides were used throughout this country. EPA proceedings in the late 1970s highlighted claims that herbicide spraying of forests in the Pacific Northwest caused miscarriages among women living there. In 1983, United States manufacture of the herbicides ended.

Swedish scientists examined the occurrence of a rare form of cancer and discovered that lumberjacks were five or six times as likely to have the disease as men in other occupations. According to the scientists' investigation, the lumberjacks were frequently exposed to dioxin-containing herbicides. Subsequent studies showed that workers exposed to the herbicides in another region of Sweden were also afflicted by the rare cancer far more often than expected. Furthermore, a more common cancer that strikes many more people was found at greater than expected frequencies in dioxin-exposed workers.

These observations of human health effects are reinforced by studies in laboratory animals that convincingly show dioxin to cause cancer, birth defects, and abortions. Results of those studies are frequently cited in the press, repeating the chemical's name and properties before the public.

However, the dread effects attributed to dioxin occur frequently in humans: About 20% of the United States population dies from cancer, a large fraction, perhaps a majority, of all pregnancies end in spontaneous abortions (often so early that the woman is unaware that she has been pregnant), and 3% of all live-born babies suffer from a serious structural birth defect. Such disasters are almost always without explanation. With no information about what caused the tragedy, it is reasonable to look at known toxic agents as possibly being responsible. The well-publicized toxicity of dioxin, in combination with information that there had been any opportunity for exposure, can lead to conclusions and contentions that it caused any of a large number of health problems. . . .

I cannot overstate the difficulties inherent in designing and executing studies to produce hard and fast answers to questions about how dioxin or any other environmental pollutant affects humans. Furthermore, the results we possess are subject to differing interpretations. Some results lead to conclusions that there are associations between exposure to dioxin and particular diseases. Other apparently contradictory studies do not find associations. In some cases, the "negative" study was too "weak" to detect the "positive" association seen in the "stronger" study. Any of a number of factors can contribute to a weak study. If the usual rate of occurrence of a disease is 1 case in 100 people, and a study examines 25 people, the rate would have to be 4 times normal to be detected. Timing can be a problem. Cancer is a disease of middle and old age and is thought to occur decades after exposure to a carcinogen. To examine a group of people 1 or 5 or 10 years after their exposure and find no excess cancer cases is not convincing evidence. Part of the confusion about possible health effects of dioxin stems from mixing together the results of weak studies that in reality do not support any conclusions with results from other, stronger studies. Other studies are flatly contradictory. For instance, the excess of a rare cancer that was reported in Swedish lumberjacks exposed to dioxin was not found in forestry workers in Finland or in herbicide sprayers in New Zealand. In those cases, detailed examination of the studies is necessary to decide why they differ and which is more likely to be correct.

To study the possible human health effects of chemicals appears, at first glance, deceptively simple. It would seem an easy task to study populations of exposed people by examining their health and medical records or by interviewing them to see if they are suffering unusual numbers or types of diseases as compared to unexposed people. Unfortunately, complications plague such analyses. Disparities in access to medical care, diagnoses, and reporting of disease among doctors and geographical areas complicate the process. Determining that a person was exposed to a particular chemical is a difficult undertaking unless the person was exposed on the job; yet workplaces are fraught with multitudes of chemicals, making it difficult to ascertain which exposure caused the ill health. Furthermore, it is almost always impossible to quantify exposures, to know which people were exposed to large amounts of the substance and which were exposed to much less. Hence epidemiology—the study of the distribution of human illneses and their causes—can be an uncertain and costly undertaking.

Furthermore, scientific knowledge of the health effects of dioxin (and other chemicals) on humans is so limited that we must rely on tests on animals to make predictions about human effects. Then scientists are saddled with the problem that measurements of dioxin's toxicity vary immensely depending on which animal is used to test it. For instance, the amount of dioxin necessary to kill one hamster would kill 5000 guinea pigs. Similarly, although there is convincing evidence that dioxin causes cancer, abortions, and birth defects in animals, the application of that information to humans is subject to dispute, especially when it has been demonstrated that some results obtained in studies on rats differ from some obtained on mice. If the tests do not agree between those relatively closely related animals, how can they be used to predict human effects?

Some proportion of the hundreds of millions of dollars allocated to scientific investigations of dioxin is being spent on animal studies. Some will sharpen our knowledge of how dioxin works at the molecular level, and may provide powerful information for making decisions about associations between the chemical and disease. At the same time, animal studies leave many people cold. We already know that dioxin causes illnesses in animals and then kills them. What more do we need to know from animals?

Perhaps the only thing certain about the results of animal tests is that the terms "uncertain" and "uncertainty" will crop up whenever they are used to make predictions about humans. In the business of predicting human risk, where a person sits or works is often predictive of what position he will take in a controversy—although this is hotly denied by participants on all sides. Manufacturers may be convinced of the safety of their products because of data or because they choose to ignore reports of their products causing adverse effects. On the other side, public interest groups may be convinced that a product is hazardous because of other facts or because of different interpretations of the same facts. Or the latter may lump all chemicals together, expecting such unattainably high standards for safety that essentially all chemicals would be considered a hazard of some kind or another.

Additional scientific inquiry may resolve some of the differing interpretations and sway some opinions, but science is slow. Moreover, the process of science itself is unsure; sometimes studies simply "don't work" and fail to provide the solid answers so desperately sought.

Science and society, in wanting to resolve difficult questions about toxic substances and other hazards, have fostered a new discipline called "risk assessment." Sold by its proponents as a rigorous undertaking that does not depend on opinions or, at least, spells out opinions, risk assessment is offered as a unique tool for making decisions on controversies such as dioxin. Its luminous promise is that it will produce an objective, quantitative estimate of dioxin's risk. That estimate can then be compared with other estimates to place dioxin in a proper perspective within a catalogue of risks that we now contemplate, regulate, or tolerate.

Risk assessment is not all that it is sometimes cracked up to be, however. It is not like addition, where the rules guarantee that everyone who knows how to carry out the manipulations will get the same answer to a given problem. Instead, different risk assessors confronted with the same problem may well produce different results. When there are many facts and opinions, some may be ignored in analyses. Some are ignored for good reason, but one person's "good reason" may be dismissed as "valueless" or "value-laden" opinion by someone else. The selection of what data to consider and what to ignore profoundly influences the course of risk assessment, as does the selection of a "mathematical model" to estimate risks when few data are available. As is well known in the risk-assessment trade, risk estimates for a single substance may vary by factors of tens, hundreds, thousands, or more depending on the data chosen and models used.

Most scientists take a tolerant view of risk assessment, warts and all. To be believed, risk assessors must lay out the data they considered, how they analyzed the data, and how they arrived at their judgments. This process of "showing your work" greatly facilitates understanding, and it is worlds better than an individual's proclamations that calculations have been done and that they show dioxin is a terrible risk or no risk at all. In such a case there is nothing to examine.

When the epidemiological studies have been completed and the animal tests finished and the risk assessors have done their job, we should have an estimate (or estimates) of the risk posed by dioxin. So what? Is the risk small enough to put up with or so great that we should spend time and money to reduce it? These are not scientific questions. In a democracy, they are questions to be answered by citizens, and in our country, we count on the courts and political system to provide answers.

Because it is more than a scientific question and problem, dioxin invites political and legal controversies, involving sick and worried people who are petitioning the government for compensation or help in reducing risk and who are suing other people for damages. In our society, we can resolve such controversies by electing people to office, influencing particular lawmakers, relying on regulatory agencies to make a decision, or going to court. In the case of dioxin, all these avenues have been employed by parties on various sides of the issue. However, all the legal and political approaches have drawn on science. Claims of damage or threats to health have been buttressed by references to epidemiological and animal studies. Estimates of possible harm have been drawn directly from risk assessment. The dioxin issue in the courts, legislatures, and regulatory agencies, so far at least, has not been divorced from scientific investigation and results.

The intertwining of legal, political, and scientific processes in efforts to resolve the controversies about dioxin has thrown policymakers and scientists together. It is an uneasy alliance, strained by suspicion on each side of the other's methods for gathering facts, synthesizing information, understanding problems, and developing solutions.

A scientific fact differs from a legal fact; the former is not accepted as "proved" unless it is verifiable by another scientist in another location and if repeated tests fail to prove it false; the latter is "proved" if a preponderance of opinion says that it took place. Many scientists can only regard with horror the idea that judges and juries can decide questions about the cause of a disease when their own approach to coming to conclusions requires years of diligent inquiry. The patience of decision-makers is strained by scientists' insistence on more time, but it is burst asunder when, at the end of their studies, scientists still remain uncertain about questions of causality.

Writing Assignments

1. Write a summary of this selection from Michael Gough's book *Dioxin, Agent Orange: The Facts*. Notice that the subtitle is *The Facts*. In your summary, indicate where Gough is presenting facts and whether he takes a position. Is he trying to be persuasive as well as informative? If so, what is he trying to persuade his readers about?

2. Gough draws a distinction between scientific and legal facts. Write a short essay that explains the significance of this distinction in making public policy about toxic substances such as dioxin.

3. Do library research to update this selection from Gough's book, which was published in 1986. In the intervening time, what new scientific evidence is available to understand the health effects of dioxin? Write a short essay that presents your findings and takes a position on what, if anything, should be done about dioxin. Take into account that you will need not only to inform but to persuade your readers.

Credits

Excerpt from CHEMISTRY FOR ENGINEERS AND SCIENTISTS by Leonard Fine and Herbert Beall, copyright © 1990 by Saunders College Publishing, reprinted by permission of Harcourt Brace & Company.

"Chemical and Physical Changes" from INTRODUCTION TO GENERAL, ORGANIC & BIOCHEMISTRY, Second Edition by Frederick A. Bettelheim and Jerry March, copyright © 1988 by Saunders College Publishing, reprinted by permission of the publisher.

Chemical Abstracts, Vol 116 (1992) and Vol 122 (1995), Copyright © by the American Chemical Society and reprinted with permission.

Reprinted with the permission of Scribner, imprint of Simon & Schuster, Inc., from THE DOUBLE HELIX by James D. Watson. Copyright © 1968 by James D. Watson.

Reprinted with permission from ISSUES IN SCIENCE AND TECHNOLOGY, Baltimore, "Science and Climate Policy: A History Lesson," Summer 1989, pp 48–53. Copyright 1989, by the National Academy of Sciences, Washington, D.C.

Pamela S. Zurer, "Scientific Whistleblower Vindicated" from CHEMICAL & ENGINEERING NEWS 69 (14), April 8, 1991. Reprinted with permission from the American Chemical Society. Copyright © 1991 American Chemical Society.

David Weaver, Chistopher Albarnese, et al. Letter to the Editor. *Cell,* (59) 1991, p. 536. Copyright © *Cell.* Reprinted by permission of *Cell.*

Excerpt from "Youth and Education", in Frederick H. Getman, THE LIFE OF IRA REMSEN, 1994. Copyright © Arno Press, reprinted by permission of Arno Press.

Watson, James D. and Francis H. C. Crick, "A Structure for Deoxyribose Nucleic Acid." Reprinted by permission from NATURE Vol. 171, pp. 737–738. Copyright © 1953 Macmillan Magazines Ltd.

Michael E. Potts and Duane R. Churchwell, "Removal of Radionuclides in Wastewaters Utilizing Potassium Terrate" *(VI), Water Environment Research* 66.2, (March/April 1994). Copyright © 1994 Water Environment Research Foundation, reprinted by permission of Water Environment Research Foundation.

Thomas Henry Huxley, "Joseph Priestly" in COLLECTED ESSAYS, Vol. 3, 1970. Copyright © Georg Verlag AG. Reprinted by permission of Georg Verlag AG.

Excerpt from W. L. Bragg, "The Structure of Some Crystals as Indicated By Their Defraction of X-Rays", from The Proceedings of the Royal Society, London, pp. 248–277 (1913). Reprinted by permission of The Royal Society, London.

Index

Acid Rain: A Plague Upon the Waters, 173–176
acid rain, 173–176
active voice, 60
analogy, 36, 37, 47, 48
annotating, 93
audience, 8, 122, 123, 161–162, 166–167

Baltimore, David, 17–29
Beilstein's Handbook of Organic Chemistry, 141
books, references to, 131–132
Bragg, William Lawrence, 114–116
Branch, G. E. K., 116–118, 121
Bray, W. C., 116–118, 121

Carson, Rachel, 167–173
Chemical Abstracts Service On Line, 139
Chemical Abstracts, 136–139, 140
    Index Guide, 137–139
Churchwell, Duane R., 100–107
competition, 14–15
computer records, 70
computer searching, *Chemical Abstracts*, 139
computers, 139
conciseness, 6–7, 9–10
controversy, scientific, 110
conventions, chemical, 47
Crick, Francis H.C., 35, 92–99
critiques, 99–100, 107–108
Current Abstracts of Chemistry, 139

data bases, computer, 136
data, 64
    raw, 64
    searching for, 141, 142
demonstrations, 45
Department of Defense, 147
Department of Energy, 147,148
Dictionary of Organic Compounds, 141
Dioxin, Agent Orange: The Facts, 177–182
dioxin, 176–182
Directory of Research Grants, 144
drafting, 133, 159

editing, 89, 134, 160
electronic notebook, 70
equations, 42, 43
error, 64, 79, 80, 87, 88
    absolute, 79
    percent, 79
    relative, 79
essay questions, 49, 50
examinations, 46–51

facts, 35, 47
first person, 60
formulas, 42, 43
funding sources, 143, 144

Getman, Frederick, 53, 54
Gmelin's Handbook of Inorganic Chemistry, 141
Gough, Michael, 177–182
grant applications, 143–150
  contacting program officers, 148, 149
  Department of Energy, 147, 148
  National Science Foundation, 145–147
  talking to past grantees, 149
  graphing, 81–86
  by computer, 86
  drawing lines, 84, 87
  simplifying numbers, 84
graphs, 42

Handbook of Chemistry and Physics, 141
Heilbron's Dictionary of Organic Compounds, 141
Hershel, William, 56
highlighting, 42, 43, 45
Huxley, Thomas Henry, 111–114
hypothesis, 74

illustrations, 42
Imanishi-Kari, Thereza, 17–18, 55
Index Chemicus, 139
index, 137, 138
  author, 137
  key word, 137, 138
  patent, 137
insecticides, 167–173
Internet, 142
interpretation, 110–121, 126, 128
isotopic lead experiment, 124–129

JANAF Thermochemical Tables, 142
journal, laboratory, 56
journals, 129–131
  articles, 135
  references to, 130, 131

Kolbe, H., 110

laboratory notebook, 55–70
  abbreviations, 57
  conclusions, 64–70
  construction, 56
  discussion, 64
  industrial, 55,56
  page numbering, 57
  plan, 58–60
  preface, 57
  procedures, 60
  purpose, 58
  results, 60–64
  table of contents, 57
  witnessing, 55, 57
laboratory report, 70–89
  calculations, 78–86, 88
  conclusions, 86–88
  discussion, 86–88
  editing, 89
  introduction, 74–76, 88
  organization, 72
  procedures, 76–78, 88
  proofreading, 89
  results, 78–86, 88
  revision, 88, 89
  preparation, 54, 55
laboratory, purpose, 52
Lavoisier, Antoine, 110–114
Le Bell, Joseph, 110
lecture notes, 45, 46
  reviewing, 45, 46, 73
  rewriting, 46
  taking, 44, 45
Lewis, Gilbert Newton, 118–121
literature review, 121–142, 151, 152, 154, 155

audience, 122,123
choosing topics, 122
lecture notes,conclusion, 128, 129, 134
exploring topic, 123
introduction, 124, 125, 134
purpose, 122, 123
references, 129–133
review, 125–128, 134
structure, 123, 124
literature searching, 134–142
logbook, 56

macroscopic level, 35
margins, 40, 45
measurements, 79
memorization, 47
Merck Index, 141
microfiche, 135
microfilm, 135
microscopic level, 35
models, 35, 36, 47
multiple choice questions, 49, 50

National Institute of Standards and Technology, 142
National Institutes of Health, 143
National Science Foundation, 143, 144
Chemistry Division, 145–147
grant applications, 145–147

O'Toole, Margot, 17–18
objectivity, 6–7, 10–11
observations, 34, 35
on-line data searching, 139, 141
Ostmann, Jr., Robert, 173–176

patents, 137
references to, 132

Potts, Michael E., 100–107
precision, 6, 8–9, 77
Priestley, Joseph, 110–114
problems, 41, 42, 48, 49
assigned, 41, 42
in examinations, 48, 49
proofreading, 89, 134, 160
proposal writing, overview, 144
public attitudes, 164–166
purpose, 122, 123

questions, test, 49–51
quizzes, 46–51

references, 74, 76, 129–133, 159
format, 130–133
Remsen, Ira, 52–54
research proposal, 143–160
anticipated results, 158, 159, 160
choosing a topic, 151
drafting, 159
editing, 160
goals, 150
importance of work, 156
literature review, 154, 155, 160
proofreading, 160
references, 159
revision, 160
structure, 152, 153
summary, 153, 154, 160
work proposed, 155–158, 160
research, chemical, 145–147
revision, 88, 89, 133, 134, 160
rounding off, 80

Science Citation Index, 139
science writing, 161–164
informative, 161–164
persuasive, 166–182

scientific articles, 91–92
discussion section, 92, 97–98
introduction, 91, 97
literature review, 91, 97
procedures, 91
references, 92
results, 91, 97
scientific honesty, 12–14, 17–29
short answer questions, 49, 51
signigicant figures, 80
Silent Spring, 167–173
skimming, 39, 42
Smalley, Richard, 34
study groups, 48
studying, 46–51
summaries, 40, 41, 44, 98–99

tables, 42, 80, 81
textbook, 37–44, 109, 121
The Chemist's Companion, 141
The Double Helix, 15–17
theories, 109
Thermophysical Properties of Matter, 141, 142
theses, references to, 132
thinking like a chemist, 48

uncertainties, 79, 80
underlining, 39, 40, 43, 88, 93

van't Hoff, Jacobus, 110
variable, 82
dependent, 82
independent, 82
visual aids, 45
voice, 60

Watson, James D., 15–17, 35, 92–99
Wiggins Compilation, 142
World Wide Web, 142

Zurer, Pamela, 29–32